THE SELLING OF NEW ZEALAND MOVIES

For Guglielmo Biraghi, Jonathan Dennis
and Jeannine Seawell

Contents

ABOUT THE AUTHOR

*L*INDSAY SHELTON BEGAN A LIFETIME fascination with the movies as a child watching Hollywood musicals in his small rural home town. After working as a journalist in Sydney and London, he returned home to join New Zealand's first television news service. In his spare time he was elected president of the Wellington Film Society. Soon afterwards he established the Wellington Film Festival which he ran for nine years until becoming marketing director of the newly established New Zealand Film Commission. For the next 22 years he introduced New Zealand movies and movie-makers at the world's biggest film festivals and markets – including 21 Cannes Film Festivals – and handled international sales of more than 60 New Zealand feature films by aspiring, and later famous, directors such as Peter Jackson, Jane Campion, Roger Donaldson, Gaylene Preston, Geoff Murphy, Vincent Ward and Lee Tamahori.

ACKNOWLEDGEMENTS

I OWE THIS BOOK TO MIKE NICOLAIDI, who suggested it and whose enthusiasm and confidence made it happen. Without his involvement and support I could never have started it, or completed it.

The idea of a book on New Zealand films was first suggested to me 25 years ago by Roger Horrocks, who has done so much for New Zealand's film culture. Two hundred and fifty films later, it's a very different book from the one he had in mind.

There are many people who are not in the narrative but whose support and friendship were a part of my years at the Film Commission. In particular I acknowledge Carol Davidson, my valued and longest-serving assistant, and Jack Ingram, Phil Langridge and Owen Lewer, who worked with me for many years in the never-ending roles of ensuring that films and tapes were delivered on time to addresses all over the world, and that thousands of pages of correspondence and contracts were stored where they could always be found. I should also acknowledge the unfailing generosity of Chris Prowse and Mladen Ivancic, who were always willing to explain financial realities which would otherwise have remained a mystery.

Until now I have never thanked the 1972 committee of the Wellington Film Society for their willingness to support the untried and unknown concept of a film festival. My special thanks go to David Lindsay, then editor of the society's magazine *Sequence*, which recorded the details of our '70s censorship battles. Like everyone involved with New Zealand films, I am also grateful to Sue May, David Gapes and Nick Grant, whose editorship of *Onfilm* magazine has provided an unequalled record of good and bad times.

I thank Ruth Harley, the chief executive of the Film Commission, for permission to use photographs and to access files from the commission's collection at the New Zealand Film Archive, and for permission to use (and revise) the commission's time line of feature films, which is the basis of the filmography in this book. Diane Pivac from the Film Archive provided authoritative input into revising, expanding and correcting the list. Ian Conrich ensured we added titles which might otherwise have been overlooked. I thank them both. Some information in the filmography has come from the time line created by Jonathan Dennis in *Film in Aotearoa New Zealand*. Another valuable resource was the landmark *New Zealand Film 1912–1996* by Helen Martin and Sam Edwards.

Frank Stark, Kristen Wineera and Lissa Mitchell all provided generous assistance when I needed to find files and information at the New Zealand Film Archive. I am grateful to many other people who answered questions when I called to check on facts which I couldn't otherwise confirm. I thank William McAloon for the story behind Gordon Walters' logo for the Film Commission. Paul Melody in Marton and Michelle Seawell in Paris were quick to provide information when I most needed it.

I am grateful to Stephen Stratford for his work on an early version of the text, Marianne Ackerman for her attractive book design, and typesetter Linda Guinness who took such care with every page. At the Wellington offices of Awa Press, it's been a pleasure to know that Sarah Bennett would always display impeccable technical knowledge and calm, cheerful efficiency. Most of all, I thank publisher Mary Varnham for her encouragement, commitment and confidence, and for being at all times unrelentingly demanding and stimulatingly constructive.

FOREWORD

ONE OF THE WONDERS OF the new century is how a small island nation at the bottom of the world emerged as a source of extraordinary film-making talent and a global film location *par excellence*. Thirty-five years ago that possibility was dreamed of by few New Zealanders, other than a handful of practitioners and film activists dedicated to shaking successive slumbering governments into action.

Yet today, following the worldwide blockbuster success of the trilogy *The Lord Of The Rings*, New Zealanders can lay claim to our first resident native film mogul. Peter Jackson has chosen to continue making films on home base and has built his own state-of-the-art studio complex in Wellington. Given the power and influence he can now wield in Hollywood and beyond, this situation may seem quixotic. But in Aotearoa New Zealand fantasy has become reality. We have tilted at windmills, and apparently won.

The tales about how the country broke through the movie barrier are many and varied – and as serious and bizarre as all the people involved. This story by Lindsay Shelton is one of the most important because Lindsay was a key figure at the front line. The making of any movie is an all-consuming, highly demanding affair, first to get the money and talent together and then to create the magic. But an even greater challenge is to then sell the film to audiences, both at home and in other countries and cultures.

The bid to establish a film industry in New Zealand was one of several passionate urgencies that arose from the ferment of the late 1960s, as the post-World War II generation of the so-called developed world sought greater personal freedoms, including freedom of creative expression. In New Zealand the desire for a palpable national identity ran strong and fierce, particularly among those working in the fledgling arts and cultural enterprises. For them, Mother England – or anywhere else for that matter – would no longer be the natural creative source, or automatically hold sway over their imaginations. This string of islands in the South Pacific and its awakening peoples were moving to centre stage – as of right.

The major breakthrough came in 1978 when, coinciding with a flurry of independent film-making by young directors, a New Zealand Film Commission was finally established to promote the production of New Zealand films. In its wake a roller-coaster ride began. In the first ten years, the production of feature films fluctuated as tax incentives first found, and then fell out of, government favour. Deregulation and opening up to foreign investment became the rubric of successive governments. Near-fatal blows to confidence were delivered by the taxman's investigations into 'special film partnerships' and by the stock-market nosedive of 1987.

With most film-makers also themselves on a fast learning curve, the quality of these early features was decidedly mixed and many failed with audiences and critics. Indeed, the continuing drama of kiwi big-screen ambition had the breathless anxiety of the 1914 Pearl White movie serial *The Perils Of Pauline*. Just about everyone involved in making movies would have identified with the heroine's desperate bids to evade attempts on her life. Meanwhile, filling the role of her dastardly guardian was the government and its funding agent the Film Commission in about equal proportion – even though the slowly evolving industry knew that without either any continuity in feature film-making would be well nigh impossible.

The first decade was proof positive how risky the game was, and how innovative and flexible and chary the Film Commission and film-makers had to be in funding and making films – and in dealing with each other. Yet an indigenous feature-film industry hung in, if at times seemingly by a thread – or, in the case of Peter Jackson and his first film, the 1988 fodder horror *Bad Taste*, by a shoestring.

A change of government from Labour to National in 1990 sped up deregulation of the total New Zealand economy. To the alarm of most film and television production houses, still at a fragile stage of development, the new administration scrapped all restrictions on foreign investment in local television, thus opening the door to 100 percent offshore ownership. It also effectively reduced the Film Commission's income by 20 percent as part of finance minister Ruth Richardson's bid to draw the line on what she saw as 'a culture of dependency'.

And yet it was during this time, the driest time of a dry political climate, that the country's film-makers broke through. In 1993 Jane Campion won the Palme d'Or at Cannes with *The Piano*. The following year Peter Jackson's *Heavenly Creatures* and Lee Tamahori's film adaptation of Alan Duff's novel *Once Were Warriors* triumphantly emerged, hauling the whole industry and its relatively short history behind them.

This was a time for rejoicing, and a golden moment for those who had doggedly supported the commission's *raison d'être*. The big success of these three films, and the general celebration of the talents that went into creating them, put into perspective the mini-alarums over film development policy, marketing, industry representation on the commission, and training. In retrospect these seemed merely prevarications, filling in time until a locally made movie hit pay dirt. That said, continuing debate about how to sustain a high-quality national film industry in a country of New Zealand's size will be ever present.

The jubilation also gave us a moment to reflect on how partnership between the two primary cultures in Aotearoa New Zealand, Maori and Pakeha, was beginning to work. Of all questions that impinged on my consciousness during my involvement with the Film Commission and Film Archive in the final decade of the twentieth century, the matter of the Maori-ness of New Zealand and its due recognition was at the forefront. Like pioneer film-maker John O'Shea, I had come to believe that the essential story of the country – not its only one but its linchpin drama with, as O'Shea put it so well, 'compulsive essence' – was anchored uneasily in the shifting sands of increasingly intimate contact between the two cultures.

O'Shea signalled the path in 1952 with his first feature film, *Broken Barrier*. But it was his remarkable *Tangata Whenua* television series – initiated by historian Michael King and directed by Barry Barclay – that in 1974 prised open the ethos that would come to occupy the nation, and provide the strongest and most potent film stories and performances well into the new century. These were the movies *Utu*, *The Quiet Earth*, *Came A Hot Friday* and *The Piano,* together with the deceptively quieter and more understated *Ruby And Rata* and *Ngati*, the antecedents of the big new-millennium success *Whale Rider*.

In this new stream of passionate outflow – enhanced by the huge box-office success of *Once Were Warriors* in New Zealand and Australia – narrow definitions of Maori and Pakeha have become less important, and may even prove untenable. The global world's remorseless ability to impress its power, and expand and speed information flow, hastens all of us on together. Extreme nationalist stances – whether linked to a nostalgic view of a colonial past or to fear of a swiftly developing multicultural future – are consumed, digested and rendered down. Authentic partnership and freedom of expression for all peoples and cultures living in Aotearoa, an old ideal, is becoming the new-century passion that finds voice and confidence among film- and video-makers in this lucky land on a shrinking globe.

If the works created by this new breed are marketed with the determination, respect, aroha and regard that so characterised Lindsay Shelton's career during the formative years of our feature-film industry, the best will certainly find their audiences. They will give sustenance both to New Zealanders and people of other lands, bringing life-enhancing laughter and tears, meanings and emotions, from which we will all benefit in the deepest sense.

MIKE NICOLAIDI
July 2005

ONE

TWO COWGIRLS
SITTING ON A CLOUD

THE FIRST FILM I REMEMBER was about two cowgirls sitting on a cloud, with funny things happening on the earth below them. Films were screened in the Civic Theatre on Broadway, the main street of Marton, my home town, population 2,900. I had no idea it might have been named after a famous street in New York. Nor could I have conceived of the idea that in less than 40 years New Yorkers would be queueing up to see movies made in my own country.

The Civic had a children's club with Saturday morning matinees which my friends attended every week. My mother didn't approve of such regular film-going. So while my friends saw every episode of the American serials, I never got to find out how the heroes and heroines escaped from the explosions or crashes or shrinking rooms.

Instead, my occasional film-going as a schoolboy at the end of the 1940s introduced me to the world of Hollywood musicals. I remember singing 'If You Knew Susie' from a movie featuring the 1930s' vaudeville star Eddie Cantor. We danced along Broadway in imitation of the Hollywood stars we'd seen at the Civic. In a country with no show business, we acted out New York stories and backstage sagas in Gavin Cooper's parents' double garage, which we set up as a theatre, with our parents and friends as the audience.

All the Hollywood films seemed to be in Technicolor. British films, though, were in black and white like the serials, and they always seemed to turn up first in Feilding, half an hour away, where my grandparents lived. My parents liked the classic Ealing comedies produced by Michael Balcon – including *Kind Hearts And Coronets* with Alec Guinness and Dennis Price, *Passport To Pimlico* with Stanley Holloway and Margaret Rutherford, and *Whisky Galore* with Basil Radford and Joan Greenwood. Critic Penelope Houston has written that these films celebrated an England closed and sheltered, a Winnie-the-Pooh land. But she also noted their grace and humour, and their genial, incompetent heroes who delighted in defying bureaucracy. We drove to Feilding to see all of them.

Occasionally there were local newsreels from the government-owned National Film Unit (I was three when it was established in 1941), but we didn't take them seriously. As Robert Allender wrote in a 1948 issue of New Zealand's literary bible *Landfall*, 'The appearance of our Prime Minister on the screen always caused loud amusement. It was all very well for Winston Churchill, but fancy our Mr Fraser in the pictures. It was all too funny.'

The Civic Theatre, built in 1904, was rented to Robert Kerridge, part of his empire of 133 cinemas, the biggest exhibition chain in New Zealand or Australia, which from 1945 was 50 percent owned by the Rank Organisation in Britain. The cinema sent everyone a monthly calendar of coming attractions. An only child, I would sometimes take it to the front porch of our house in Signal Street and shout announcements about coming attractions through the horn of my old gramophone.

When I was 13 I was sent to boarding school in the nearby town of Wanganui, and film-going became less frequent. The Collegiate School allowed only occasional outings to films and tended to favour British productions. We were briefly taught that British films showed reality in an unreal way, while Hollywood showed fantasy as if it were real. We had to write an essay analysing David Lean's *The Sound Barrier*, a story of the British upper crust. But this was the only time I remember film entering the curriculum: it wasn't a subject to be taken seriously, like English or music or Latin or physics. There seemed no possibility it could become a career.

CinemaScope arrived in 1953. Heading for a family holiday in the Bay of Islands, we stopped in Auckland and saw *The Robe* with Richard Burton and Victor Mature on the unbelievably wide screen that had been installed in the Civic Theatre. Elia Kazan's *East Of Eden* and Nicholas Ray's *Rebel Without A Cause*, the first films starring the electric James Dean, arrived a year or two later, also in the new format.

When I moved to Wellington in 1956 to work on *The Dominion* newspaper and study at Victoria University, I discovered double bills of older films at the Roxy and Princess movie houses. Each offered four sessions a day with a rich all-American diet of *film noir*, although we didn't know the term yet. There were gangster films and science-fiction films, and more musicals, including overlooked ones such as Doris Day and Kirk Douglas in *Young Man With A Horn*, made in 1951 by Hungarian-born Michael Curtiz, who had directed the now-legendary *Casablanca* nine years earlier. In Wellington in the '50s the movies were where you went with your girlfriend on Saturday nights. Even in the gigantic Majestic, which showed the Doris Day and Rock Hudson comedies, you had to book because the two thousand seats were always sold out. I loved the enormity of the Majestic and its huge domed ceiling. I loved the dark-panelled, thickly carpeted public areas of the Regent, and the light-shows on the curving proscenium arches at the art deco Plaza.

This was film-going as New Zealand's most popular leisure activity, with pre-television attendances peaking at 40 million ticket sales a year in a population of only 2.4 million. There was also occasional live theatre, pre-eminently the touring New Zealand Players headed by Richard and Edith Campion. In Wellington the theatrical high point was the annual university revue *Extravaganza*, produced in the Opera House by a young law graduate from Taranaki named Bill Sheat.

The only film magazine I knew about was *Films and Filming*, a British monthly which seemed to headline sex on the cover of every issue. The only critic I read was Russian-born Catherine de la Roche, who wrote in *The Dominion*. De la Roche had emigrated from London, where she had been a contributor to *Sight and Sound*. I'd never heard of that magazine. And I'd never heard of film societies.

Although we never saw a locally made film, the absence didn't seem strange. We didn't think of New Zealand as having any stories worth telling, or any places worth showing. Gordon Mirams, the much-admired film critic of the *New Zealand Listener*, wrote that if there were such a thing as a 'New Zealand culture', it was to a large extent the creation of Hollywood.

A small group of people were trying to change this, and I had my first contact with them in 1957. *The Dominion* assigned me to write a story about a road-safety film being made at MacKays Crossing north of Paekakariki. It was a night shoot with bright lights, people working briskly in the shadows and a train waiting for its cue. The film's producer and director was John O'Shea whose 1952 *Broken Barrier* (co-directed with Roger Mirams) had been the first feature film made in New Zealand in twelve years. After completing it, with no money for any more full-length films, he and his colleagues had started making sponsored shorts, including 25 on road safety, and a series of documentaries about the Pacific sponsored by New Zealand airline TEAL (Tasman Empire Airways Limited).

Tony Williams, the cameraman on the Paekakariki film (and later a very successful producer of television commercials), would many years afterwards, in an article in film magazine *Illusions*, recall it as a time when New Zealanders who worked in films were considered mad, batty or just weird. At this first encounter I don't remember thinking anyone was mad, but I sensed a mood of excitement and concentrated efficiency. After my *Dominion* article was published, O'Shea wrote a letter thanking me, but I wasn't encouraged by the paper to pursue the connection with Wellington's tiny film-making community. I was promoted to Hutt Valley reporter, driving the office's only car to Lower Hutt to cover the trial of schoolchildren charged with under-age sex on the banks of the Hutt River. No one could understand how secondary schoolchildren could know about sex or want to participate in it.

Towards the end of the '50s I got a spare-time job writing and editing a free weekly suburban paper called *Newtown News*, taking over the job from the poet Louis Johnson. I started a film page to help the two Newtown cinemas, whose managers gave me pressbooks from the Hollywood studios with ready-to-use stories about the stars. I hung out with the manager of the Ascot and his wife, and watched movies from their projection box. Nicholas Ray's *Party Girl* with Robert Taylor and Cyd Charisse, wearing red, seemed extraordinarily sophisticated.

In 1960 I was recruited to be a sub-editor on *The Sydney Morning Herald*, paid more as a 21-year-old journalist in Australia than my father Norman was earning as a member of parliament at home. Sydney had three cinemas specialising in foreign-language films with English subtitles. The first foreign film I saw was the sexy French thriller *Plein Soleil* with Alain Delon. The second was Mauro Bolognini's black and white mood piece *Notte Brava*, with young Italian stars wearing wet T-shirts and riding motorbikes, and a script by Pier Paolo Pasolini, a poet who wasn't yet a director. Such films offered an alluring openness and style I hadn't experienced before.

Sydney also had three television stations, when in New Zealand television was broadcast only in Auckland. (It would come to Wellington and Christchurch in 1961 and Dunedin in 1962.) The schedules included old black and white movies at midday, although there were only enough to support a three-month cycle before they started to be repeated. My mother came to visit and watched the television test pattern; she was amazed when the programmes began and there were moving images.

There was much more to amaze us. In New Zealand, for example, broadcasters were required to have pseudo-British accents; on television and radio in Sydney everyone spoke with a distinct Australian twang.

In 1962 my English-born wife Vivien and I moved to London, where I took up a position as a correspondent for the Sydney *Sun-Herald*. In a new country I wondered if I could make a change of career. I applied for a production trainee's position on the BBC's second channel, which was about to be established. After a year I was offered a job, but by then Vivien and I had decided to leave our attic in South Clapham and return to the southern hemisphere.

We intended to go back to Sydney, but when our boat arrived in Wellington one of the journalists who came out to interview the new arrivals was Jim Siers, with whom I had worked on *The Dominion*. He told me about New Zealand's new television news service, which he said was looking for staff. I went to see the editor, Ben Coury, and soon joined Doug Eckhoff and a small team pioneering the seven o'clock evening television news for the government-owned New Zealand Broadcasting Corporation. We wrote the bulletins in offices in Broadcasting House behind Parliament Buildings, and then walked five blocks across town to oversee the telecast in a converted radio station in Waring Taylor Street.

When I arrived, the news seemed to consist of nothing but a reader sitting in front of a camera. Through my Fleet Street contacts I got wire photos to illustrate overseas items. A film newsreel followed the 'spoken' news, and soon we started to merge moving pictures into the bulletin. Recorded newscasts from two American networks were flown to New Zealand every day and we re-broadcast American correspondents' filmed reports from the Vietnam war – in spite of behind-the-scenes pressure from politicians who wanted less time allocated to such disturbing images. These were the days of 'all the way with LBJ'.

The news became the top-rating show on the country's only television channel. Every year there was some new technique or technology to bring to our large audience. We created a dream routine of eight days on, six days off, later improving it to seven 12-hour working days followed by seven non-working days.

Once my children Adam and Kate started school, all the spare time gave me an incentive to seek some extra activity. I tried directing shows for Wellington Repertory, but the world of amateur theatre was full of frustrations. Then in 1969 I saw an advertisement announcing the annual meeting of the Wellington Film Society. Because I put my hand up and asked questions, I was co-opted to the committee. A year later I was elected president, and outgoing president Ron Ritchie asked me to take over the programming of all 50 film societies through the Federation of Film Societies. The job had been done for many years by his wife Margaret, who had died. Ron gave me boxes of her meticulously sorted files of correspondence and set off for a trip to England.

The Wellington Film Society had been established in 1945, with the *Listener* critic Gordon Mirams as its first president. Mirams' friend John O'Shea edited the monthly magazine and contributed articles and reviews. 'We were mad about films,' he later wrote in his memoir *Don't Let It Get You*. Within two years he and Mirams had established a national federation, to import annual programmes of exceptional and acclaimed films as an alternative to the fare in regular cinemas.

Tahu Shankland, later to become an influential television programmer, was also on the committee, as was Walter Harris, an educationalist from Christchurch. Harris had established the Education Department's National Film Library, using it as a base to introduce film into school curricula and to collect prints of early films. Forty years later these prints would become the basis of the new New Zealand Film Archive's collection.

Every year the society presented nine or ten films in the lecture hall of the Central Library, screening 16-millimetre prints to a membership of around 300. Its annual brochure was a plain affair, and its programming mainly film classics, with *Battleship Potemkin* a regular favourite. From my film-going in Sydney and London I knew there were many great new films that weren't reaching New Zealand. I decided to try to bring them into the country for the film society circuit.

In 1970 my family and I went to London to visit my wife's parents. I spent much of the holiday introducing myself to British distributors, most productively Charles and Kitty Cooper of Contemporary Films, who had British rights to a large catalogue of distinguished foreign films and were willing to get permission for the films to screen in other countries. There was an enormous range of films screening in London, with cutting-edge programming led by Derek Hill's New Cinema Club. Hill's promotions made much use of the censorship issue: as a members-only club he could show films which had been banned by the British censors.

I returned home with the New Zealand rights to a programme of critically lauded new films just released in London cinemas. The film censor, Doug McIntosh, refused to allow film societies to screen Andy Warhol's *Flesh*, starring 19-year-old Joe Dallesandro as a male prostitute, with nudity and sex scenes that had caused an uproar in Britain. However he passed all the others, including *Scorpio Rising*, a film made in 1962 by the American underground director Kenneth Anger. Anger had intercut homo-erotic scenes of leather-clad bikers with clips from old films about the life of Christ, and set the whole thing to an accompaniment of songs by Ray Charles, Bobby Vinton and Elvis Presley. On the same bill we offered Anger's *Invocation Of My Demon Brother* (1969), with a soundtrack by the Rolling Stones' Mick Jagger.

Our brochures for the revitalised Wellington Film Society owed much to the New Cinema Club, with quotes from reviews by famous critics and covers promising forbidden pleasures. The number of members soon doubled, trebled and eventually quadrupled. The 1971 cover from *Duet For Cannibals*, a feature made in Sweden by the American author Susan Sontag, showed a black-clad woman holding up a very small naked man. In 1972 we used a photo of a naked woman who faced away from the camera while a man (French actor Michel Piccoli, I think) poured something down her back. Some new members told us they had joined in hopes of finding out what it was that he had poured.

We enlivened the society's monthly magazine with recommendations for films which no one had previously realised were important. Playwright Michael Heath and his friend Peter Herbert wrote entertainingly and persuasively about titles the rest of us had never heard of. These films mostly appeared on late-night television, or were rushed in and out of cinemas when big-name attractions weren't available.

On our second application, the censor approved a 16-millimetre print of *I Am Curious-Yellow*. Vilgot Sjoman's 1967 Swedish feature had become notorious in the United States, where a well-publicised Supreme Court battle had been necessary before it was cleared of obscenity charges. We advertised the film with the same image that had been used for the American campaign. Wellington Film Society membership peaked at 2,700, making it the biggest in the country and one of the biggest in the world. The membership secretary, a volunteer, resigned when the amount of mail became more than she and her father could cope with on their kitchen table.

Film societies had a technical limitation: they could screen only 16-millimetre prints. Many more titles were available from my contacts in London, but only on professional 35-millimetre prints suitable for commercial cinemas. It didn't seem possible for film societies to move into cinemas, but it did seem possible to create a film festival and present 35-millimetre programmes in someone else's. Early in 1972 I flew to Australia to learn about the Sydney Film Festival. I discovered from its director, David Stratton, that some of the prints from his programme would be available to screen in Wellington. So with backing from the film society committee, and selfless help from volunteer secretary Rosemary Hope (later to become the first full-time film society employee), I decided to create the Wellington Film Festival.

1ST WELLINGTON
FILM FESTIVAL
1972

TWO

MISS BARTLETT'S SUITCASE

*N*EW ZEALAND'S FILM DISTRIBUTORS and exhibitors told me it would be useless to start a film festival. They were smugly confident that New Zealanders wanted movies only from the mainstream American and British suppliers – the films which were so profitable for their businesses. Wellington cinema managers had no faith in the idea either. They all said no to my request to book a theatre for a week – all except Merv and Carol Kisby who ran the independent Paramount Cinema in Courtenay Place. They agreed to give us a chance.

In the first film festival in 1972 we presented a seven-day programme, including New Zealand premieres of seven features – two by the great French directors Eric Rohmer and Louis Malle, two by Italians Gillo Pontecorvo and Mauro Bolognini, and one from Spain – Bunuel's *Tristana,* starring Catherine Deneuve. From the first year we were able to prove that there was a local audience for such films. Five thousand tickets were sold. Within five years, attendances grew to over 20,000. By 2004 the festival would sell a record 74,000 seats.

I wanted New Zealand films for the first festival, but couldn't find any. However, the festival's growth was to be paralleled by a growing campaign to get government support for New Zealand film-making. At the government-financed Arts Council, with its mission to develop New Zealand culture, Bill Sheat had become chair and 33-year-old journalist Mike Nicolaidi, home after three years as the Press Association's London correspondent, the first New Zealand-born director. Both had been involved in film-making, Sheat as a producer, and a friend and supporter of John O'Shea, and Nicolaidi as a writer and director of short films. Both wanted to find ways of financing local productions.

They commissioned a report from John Reid, founder of the Canterbury University Film Society, chair of the film society federation, and an actor who had studied with the influential and autocratic Christchurch director and author Ngaio Marsh. Reid's report, published at the end of 1972, said independent film-makers had to struggle against double jeopardy: the government's National Film Unit, which controlled the tools and hardware needed for film-making and didn't want assist private film-makers, and the powerful state-financed New Zealand Broadcasting Corporation, which had a television monopoly and no policy to foster any local production except its own. The report fuelled a growing debate about the need for a film commission.

It would be 1974 before the festival showed its first New Zealand films, two shorts: a documentary on Maori artist Ralph Hotere, the sixth film directed at the Film Unit by American-born Sam Pillsbury; and *Paradise Transport*, a surrealist film showing 'flashes from the past, present and future of any South Pacific colony', made by Richard Phelps with $3,000 from the Arts Council. By this time an Auckland group calling itself the Alternative Cinema cooperative had joined the push for a local film industry, publishing a crusading magazine and scheduling film screenings and seminars at rented premises.

The 1975 film festival included our first New Zealand feature – the experimental *Test Pictures: Eleven Vignettes From A Relationship*. Geoff Steven, one of the founders of Alternative Cinema, had shot the film two years earlier, co-directing it with Denis Taylor and Philip Dadson. They had run out of money after spending their $14,000 budget. The film society paid $500 so a release print could be made. Geoff stayed with me when he came to Wellington to introduce the world premiere of the low-key story about alternative lifestylers in a small enclosed community. 'Alienation and observation are the chosen style,' noted critic Peter Harcourt in the programme.

Members of parliament who came to the premiere were uncomfortable: they couldn't find any connection between familiar Hollywood genres and the lengthy uninterrupted takes of the local production. Auckland University film lecturer Roger Horrocks wrote in the student newspaper *Craccum*: 'It is so uncompromising – or so foolhardy if you like – that I don't think the group has much chance of retrieving their money.'

As a supporting film I selected *Mururoa 73*, a documentary which showed the French navy capturing a boat sent by the New Zealand government to protest against French nuclear testing in the Pacific. The subject probably contributed to the discomfort of some of our guests. The director, Alister Barry, would go on to contribute to the discomfort of many of their successors as well, with a television series on New Zealand's role in Vietnam, another documentary on anti-nuclear policies, and then two impeccably researched features on the policies of the

New Right (*Someone Else's Country*, 1996) and unemployment (*In A Land Of Plenty*, 2002).

Early in 1975, the Arts Council had sent a report to the Labour government recommending the establishment of an annual film production fund of not less than $300,000. The report, written by Mike Nicolaidi, said the aim must be 'no less than to establish a viable motion picture industry' which would 'reflect the New Zealand way of life with truth and artistry, showing New Zealand to New Zealanders and to the world'. But not enough political support could be found to make anything happen.

In 1976 the Wellington Film Festival expanded to two weeks, with 30 new features, none of them from New Zealand. I noted in the programme: 'There is no doubt that successive New Zealand governments by their negligence must bear responsibility for the fact that our own directors cannot work in their chosen field, and thus our audiences continue to see only images of places far from here.'

The programme included an Iranian film, *Prince Ehtejab*, in which a dying nobleman recalled his extravagant ancestors. When a decade later I met the director, Bahman Farmanara, he had left Iran and settled in Canada, where he had become an influential film distributor. I told him I had wanted to keep his film because it was so beautiful. He told me he wished I had, because all the materials had since been destroyed.

Also in the festival was a cryptic French film, *Nathalie Granger*, directed by the prolific author Marguerite Duras. Its star was Jeanne Moreau, who gave a hypnotic performance. The poster was a close-up of Moreau. I framed it and hung it in a room at home.

In 1977 we found a second New Zealand feature. *Landfall*, written and directed by Paul Maunder at the National Film Unit (where it was edited by Sam Pillsbury), had been made as a television movie for the Broadcasting Corporation, and had won first prize at the Abu Shiraz Young Film-makers' Festival in Iran two years earlier. Local television programmers seemed reluctant to broadcast it, and even more reluctant to let us screen it. Negotiations took more than twelve months.

One of the actors in *Landfall* was a young director from the Film Unit named Sam Neill. Neill's talent for performing had been discovered when he had played a priest in *Ashes*, a short film directed by Maori film-maker Barry Barclay and produced by John O'Shea. Another *Landfall* actor was film society member Jonathan Dennis, who regularly attended the film festival, where he always sat in the centre of the front row. Reviewing the festival on a radio programme called *The Critics*, Jonathan gave a glimpse of the forthright opinions for which he would later become famous on Radio New Zealand's *Film Show*: 'This year was a good year. Only seven films were inconsequential and pretty execrable, three unsatisfactory or merely feeble, eight mildly interesting, eight worth seeing and four were superior achievements.'

Films occupied almost all my spare time when I wasn't working in television. For five years I was simultaneously director of the film festival, president of the film society and programmer of the Federation of Film Societies. When I was invited to review films on the ZB radio network, I couldn't resist adding something extra. Once a week I would drive to Broadcasting House to relay five minutes of comments about current releases, with plenty of opportunity to bewail the lack of locally made movies.

My film-festival credentials also helped get me a job as film critic on a weekly prime-time television show called *The Media*. *Dominion* television reviewer Dave Smith wrote that I had become an all-round entertainer. But the unscripted and unrehearsed show wasn't extended for a second year. My short life as a television star was over.

My film life, though, kept getting more complicated, largely because of conflict with the film censor. A middle-of-the-road New Zealander whom you could meet any Friday night at his Cuba Street pub, Doug McIntosh was in no doubt about the need to ban films he considered offensive, and to cut words or images which were, in his judgement, indecent – regardless of the effect such cuts would have on the film. One of his predecessors had been Gordon Mirams, who had gained a reputation as a liberal censor. Working with the same legislation – in force since 1916 – McIntosh cut films by the score, including those destined for film societies and the film festival.

When the censor restricted *Test Pictures* to audiences of 18 and over, Geoff Steven said New Zealand film-makers would have to be prepared to fight for the merits and relevance of their films against the archaic and unnecessary law. But he was lucky – his film hadn't been cut or banned.

From the Swedish Film Institute, I brought in an expensive new colour print of *The Bookseller Who Gave Up Bathing* to show to film societies. It was a charming and inoffensive period story about a bachelor who, late in life, marries a woman with a past he doesn't know about. Her trousseau includes a large cabinet. The newly-weds are idyllically happy until one day, when his wife is out of town, the man opens a drawer in the cabinet and finds pornographic postcards featuring her. McIntosh told me he couldn't allow a close-up of the postcards, because they were pornographic. This was the point of the story, I told him. The censor didn't care. He removed the shot of the postcards. The film was changed to the story of an idyllic marriage which was ruined when the husband opened the drawer of a cabinet.

In 1975 the film festival included two features from the renaissance of Australian film-making: Peter Weir's *The Cars That Ate Paris* and Michael Thornhill's *Between Wars.* In the Thornhill movie, the word 'fuck' was spoken several times. 'Indecent,' pronounced the censor, who said he had to cut the print to remove the words because they should not be heard in a public place. 'But it's not our print,' I told him. 'You can't cut a print we don't own.' To my surprise he agreed to stick black tape over the soundtrack so the words would not be heard and the print wouldn't be cut.

When Michael Thornhill arrived in Wellington as a festival guest he was incensed to learn that his freedom of expression had been interfered with, especially as his film had been shown in Australia without cuts. He climbed the unsteady ladder to the Paramount's projection box and peeled the censor's black tape off the soundtrack. When he introduced the screening of his film, he announced what he had done. The audience applauded enthusiastically. Some people cheered. The four-letter words were heard, and no one was offended except the censor. No more favours for the film festival, he said. It had been a very small favour: he had cut dialogue from eleven other films.

Actions as well as words offended McIntosh. He cut a film, based on a Strindberg play, which the German Embassy had lent to film societies. 'Copulation' was against public order and decency, he explained, as he removed five seconds of a close-up showing a naked couple embracing, filmed discreetly from above the waist.

A young lawyer, David Gascoigne, would help reform the 60-year-old censorship laws. Gascoigne, who became chair of the Federation of Film Societies, had organised appeals against three of the censor's decisions. The first had been when the censor demanded cuts in *I Am Curious–Yellow.* In spite of having playwright Bruce Mason and history professor Peter Munz as expert witnesses, we had lost that appeal. We lost the two other appeals as well.

Then in 1975 two things happened: the Labour government introduced cautious revisions to the censorship law, and a private member's bill, drawing substantially on material written by Gascoigne, was introduced by MP Jonathan Hunt. But it would take two more years, and a change of government, before new censorship regulations became law.

Alan Highet, the minister for arts, recreation and sports who wrangled the new law through parliament, was a Dunedin-born accountant. He had become member of parliament for Remuera in Auckland in 1966 after moving north from Wellington, where he had chaired the city's first arts festival and helped establish the Youth Orchestra. Highet was one of the only members of Robert Muldoon's National cabinet with any interest in cinema. In this he was encouraged by his wife, the painter Shona MacFarlane, who had been a board member of the Arts Council when it started making grants for film-making. She once told me that she never let Alan go to sleep without first whispering, 'No censorship, no censorship, no censorship.'

The counter campaign, in favour of tougher censorship, was led by Patricia Bartlett, a former nun. When Gascoigne and I turned up to state our views to parliament's select committee, Miss Bartlett arrived carrying a large suitcase. When she opened it, piles of *Penthouse* magazines fell out on to the committee table. She seemed fascinated by the magazines as she passed them around to the parliamentarians, apparently intending the nation's elected representatives to be offended by the images of naked women. They remained impassive as they turned the pages and glimpsed the centre spreads.

Miss Bartlett was also obsessed with language. McIntosh had banned Bob Fosse's film *Lenny*, in which a young Dustin Hoffman played the comedian Lenny Bruce, whose routines were heavily punctuated with four-letter words. When an appeal led to the film being released uncut, it was, she said, 'an all-time language low' which would allow 'this word to be heard by thousands of sober 20-year-olds in a picture theatre.'

The National government's legislation, purloining the wording of Hunt's bill, required the censor to consider the nature of a film as a whole, as well as the nature of the audience. The censor had made 1,093 cuts in films during 1976; a year after the new law was passed there were only 202. For film festivals and film societies, if not for everyday film-goers, censorship became a thing of the past. To approve films without cuts, the new regime used certificates such as 'Only for film societies' or 'Only for film festival audiences'. The certificates were a useful selling tool, helping attract more and more people to see films they might never be allowed to see anywhere else.

Doug McIntosh didn't have to cope with the new law. He died on Christmas Day 1976, four months before it came into effect.

Having lost the battle against censorship reforms, Patricia Bartlett renewed her campaign when legislation to create a film commission was introduced. She would succeed in persuading the government to add a sub-clause insisting that decision-makers had regard to 'standards that are generally acceptable in the community'. The sub-clause wouldn't restrict any of the emerging film-makers from telling the stories they wanted to tell, as the Highets would soon discover.

THREE

SLEEPING DOGS
AND WAKING DREAMS

*T*he concept of a film commission had been first proposed by film-maker John O'Shea at an arts conference in 1970, though he didn't use those words. He delivered a keynote speech saying that New Zealand should have a national screen organisation. The conference responded with a resolution recommending the government 'foster creative activity in films for cinema and television, and create an archive for film'.

O'Shea had been campaigning for a local film industry since his film society days in the 1940s. At a time when New Zealand had 500 cinemas but little film production except occasional travelogues, he had joined Gordon Mirams in arguing that the country should be establishing a national identity through film. Mirams' reviews, said O'Shea, persuaded New Zealanders that films were not merely sideshows but a pervasive factor in forming the community's habits and beliefs. The pair's campaign had encouraged the Labour government to establish the National Film Unit. But in those days campaigners were thinking of the need for local documentaries rather than anything more ambitious.

In the 1950s, after working briefly with Mirams as assistant film censor, O'Shea had emerged as New Zealand's most active and influential film-maker, second only to English-born Aucklander Rudall Hayward, who had become a prolific maker of New Zealand films in the 1920s when local film-going was dominated by American westerns. Hayward had been 21 when he directed his first silent feature in 1922. It was only the fourth feature film made by a New Zealander.

My Lady Of The Cave had been followed three years later by *Rewi's Last Stand*, about the Waikato land wars. Then came *The Te Kooti Trail* (1927) and *Bush Cinderella* (1928), together with 23 two-reel comedies which he made with local casts and locations while travelling from town to town. When in 1940 he remade *Rewi's Last Stand* with sound, British director John Grierson, in the opening-night audience, famously remarked that it was 'more important that New Zealanders should have produced that film than that they should see a hundred films from Hollywood'.

Hayward had then supported himself by working as a newsreel cameraman and making documentaries and travel films with his second wife Ramai, the star of the second *Rewi's Last Stand*. The couple worked for a time in England and Australia, and also filmed in China and Albania. Hayward didn't direct another New Zealand feature for more than 30 years, when he and Ramai made *To Love A Maori*, a story about love and racial discrimination, released in 1972.

The couple bypassed the established distribution systems and went from town to town screening and promoting their film themselves, just as Hayward had done with his two-reel comedies more than 40 years earlier. Ramai Hayward told me that when they arrived in each town they worked the phones, using two copies of the local phone book and calling every person with a Maori name to tell them about their film. The campaign worked well in Wellington, where the Paramount was packed for opening night. But the demands were exhausting: two years later they were still on the road promoting the film when Rudall Hayward, aged 73, died in Dunedin.

Twenty years younger than Hayward, O'Shea had co-founded Pacific Films in Wellington with Gordon Mirams' younger brother, Roger. The company had survived against the odds since 1952, when the pair had co-directed *Broken Barrier,* about a love affair between a Pakeha man and a Maori woman – the first local feature since Hayward's 1940 production.

'Make no mistake,' said their advertising. 'This is not a documentary, a featurette or a tourist travelogue.'

In the '60s O'Shea had directed New Zealand's next two features, the thriller *Runaway* and the musical *Don't Let It Get You*, which starred just about every New Zealand musician from Howard Morrison to Kiri Te Kanawa.

With the exception of a Film Unit documentary on the Commonwealth Games, O'Shea's three movies were the only features made in New Zealand during three decades. All were made with private finance. All are now national treasures, as are the more primitive productions of Rudall Hayward. But when they were first released they received little recognition and earned little from the box office.

O'Shea kept Pacific Films afloat by producing rugby newsreels, occasional television commercials, and sponsored documentaries – including all those road-safety films. He never directed another feature, but as an influential although frequently frustrated producer he employed many of the people who would make it possible for the film industry to be reborn. He was also an early influence on Jane Campion (who appeared at the age of five in an O'Shea documentary promoting milk) and gave her enthusiastic encouragement from the beginning of her career as a film-maker.

Some of his most creative productions were made for television during a four-year period in the '70s which Roger Horrocks has described as a golden age. Tahu Shankland, who had become television's controller of programmes, reversed the usual in-house production policies and commissioned documentaries from independent film-makers.

The most memorable of these films were made at Pacific Films, produced by O'Shea, directed by Tony Williams (who had been director of photography on *Runaway* and *Don't Let It Get You*) and written by my film society collaborator Michael Heath. They included *Getting Together*, about clubs and societies; *The Unbelievable Glory Of The Human Voice*, about songs and singers, featuring Michael in drag; and *The Day We Landed On The Most Perfect Planet In The Universe*, in which playwright Robert Lord joined the team to create what O'Shea would later describe as 'a strange juvenile hymn to freedom'.

In 1974 O'Shea also produced *Tangata Whenua*, six television documentaries on Maori life and culture, directed by Barry Barclay and researched and written by journalist-turned-historian Michael King. Barclay has written that the series was completed at a time when Maoridom had not been portrayed on television at all, and when the Maori language seemed 'something unspeakably foreign yet of our own country'.

Barclay, who grew up on a hill-country farm, had trained in Australia for the Roman Catholic priesthood. After he left the Redemptorist monastery he started making trade films, before joining O'Shea for a film-making partnership that would last twenty years.

The early '70s also saw the film debut of former schoolteacher Geoff Murphy and his Acme Sausage Company. Murphy's 30-minute *Tankbusters* was a speedily entertaining story of four students deciding to rob a safe at a university. It was notable for the performance by Murphy's brother-in-law Bruno Lawrence. Lawrence, a brilliant musician and actor, had founded the nomadic group Blerta, which made four tours of New Zealand presenting music, plays, and silent films directed by Murphy. *Tankbusters* was briefly released on a double bill with a Marx Brothers comedy. But like the experimental *Test Pictures* and *Landfall,* it was seen by very few.

The experience of Hayward and O'Shea seemed to show it wasn't possible to sustain feature film-making in New Zealand. However a new generation didn't want to give up: in 1977 three local features were completed and released, all by young directors making their first full-length films.

First out was a double bill directed by Geoff Murphy. The main feature, an anarchic film called *Wildman,* starred Bruno Lawrence, who co-wrote it and produced it with Murphy's brother Roy, using material they had shot for a Blerta television series.

'*Wildman* was only made because it was so cheap it didn't need any investors,' Murphy would tell the Australian film magazine *Cinema Papers.*

The supporting film was *Dagg Day Afternoon,* with comedian John Clarke as farm yokel Fred Dagg and his seven brothers. The industry was tiny: one person, John Barnett, executive-produced the first film and produced the second.

The second feature, *Off The Edge*, was an adventure documentary about skiing and hang-gliding in New Zealand's Southern Alps, produced and directed by Michael Firth. It was released by independent distributor Barrie Everard.

But the most successful of the three releases was *Sleeping Dogs.* Its director, Roger Donaldson, was an Australian-born photographer who had settled in Auckland in 1968 and begun making short films a year later. His first dramatic film had been *Derek*, a 40-minute drama he had co-directed in 1974 with Ian Mune and architect David Mitchell.

Mune had returned home – 'where I can talk my own language,' he would explain – after three years working in theatres in Wales and other parts of Great Britain. He had directed and acted at Downstage in Wellington and the Mercury in Auckland, and worked in television, before meeting Donaldson and forming Aardvark-Mune Productions. In *Derek* he took the leading role of a man who realises his life is going nowhere.

'We shot the film in a weekend. We borrowed the equipment from a friend. We had no money but somehow we scraped by,' he would recall.

Next, with finance from the Arts Council, the Education Department, the Broadcasting Corporation, the National Film Unit and themselves, they made a television series they called *Winners And Losers.* They became pioneer film marketers, taking the seven half-hour films to two international television markets in 1976 and travelling through Europe and the United States making sales to broadcasters. When I started attending the same markets three years later, buyers spoke warmly of their experiences with the two New Zealanders. 'Where is Mr Mune? Where is Mr Donaldson?' asked several Scandinavian women who were buyers for state television stations.

Merchant banker Larry Parr helped find the money for *Sleeping Dogs*, which was based on New Zealand writer C.K. Stead's novel *Smith's Dream*: $150,000 as an interest-free loan from Broadbank, a privately owned merchant bank; money from the government-established Development Finance Corporation; a $100,000 underwrite from the Arts Council; and a small amount from television in return for exclusive broadcast rights for more than 20 years.

Donaldson – who would later direct such top Hollywood stars as Kevin Costner, Mel Gibson, Robin Williams and Tom Cruise – cast Sam Neill as a man-alone hermit living on an island off the Coromandel peninsula, who returns to the mainland to become reluctantly involved in fighting a repressive dictatorship and its American supporters.

Mune, who had written the screenplay with poet Arthur Baysting, played the villain. The cast also included Hollywood actor Warren Oates and Wellington actor Donna Akersten as his girlfriend.

At a preview in Wellington's Kings Theatre, Mune sat on the edge of the circle balcony and lectured us about the importance of the film for all New Zealanders. He asked his audience to talk about it and encourage people to see it and form their own opinions. It was the first New Zealand film most of us had seen, and the first time we had been shocked into recognising that local production could be important.

Sleeping Dogs was seen by more than 250,000 New Zealanders. It was the first local feature to be accepted by large numbers of local film-goers and to prove to them that New Zealand could make films which were just as entertaining as the endless stream from Hollywood. Wellington critic Peter Harcourt called it 'an accurate, honest, funny, sensitive and un-retouched picture of how many people in this country talk and behave'. The film, he said, had overcome 'kiwi cringe', the belief that anything imported was better than the New Zealand equivalent.

After the unprecedented three local releases in one year, a newly formed industry lobby group, the New Zealand Academy of Motion Pictures, was able to claim that one in five New Zealanders had seen a New Zealand-made feature. Two Wellington cinemas and three Christchurch cinemas were showing long-running New Zealand films simultaneously – there had never been a time like this. With politicians obsessed by foreign exchange, Bill Sheat pointed out that the local releases had saved a million dollars in overseas funds.

The mini-explosion, though, did not create a sustainable industry. While some production costs may have been repaid, earnings weren't sufficient to pay for any more films. The only way New Zealand could have an ongoing film industry, everyone argued, would be with support from the state. The argument was put most vigorously in 1977 at a university arts conference where film-makers from all over the country signed a petition to parliament.

Finally, boosted by the popularity of the new films, most of all by *Sleeping Dogs*, the seven years of campaigning and reports and recommendations bore fruit. In the last weeks of 1977 Alan Highet, having persuaded his government colleagues to support him, introduced legislation to establish a film commission.

Highet had not only fought hard in his caucus, he'd also had to resist pressure from Hollywood. During a visit he and Shona MacFarlane had made to Los Angeles they had been entertained by Denis Stanfill, chairman and president of 20th Century Fox, who had told them firmly there was no need for a local film industry because Hollywood could supply all the needs of New Zealand film-goers. Shona noted that Stanfill had earned a bonus of US$1.5 million because of his company's international success.

In the programme of the 1978 Wellington Film Festival, Alan Highet predicted the country could confidently look forward to a steadily increasing number of good New Zealand films. By then Tony Williams' first feature, the gentle love story *Solo*, had joined the list of local releases. Williams had directed the advertising campaign which had helped the National Party become the government. He had given a message to party leader Robert Muldoon about the need for a film industry. Perhaps Muldoon had heeded his words.

'It is no coincidence that the Film Festival has grown at the same time as the rebirth of the New Zealand film industry,' I wrote in the festival programme. 'I hope the festival will be a means of intensifying the debate which is now beginning about what kind of films our country should be producing. Central to this debate is whether film-makers should be aiming at a local or an international market, and whether budgets should be high or low.'

That debate would never end. It would be just as vigorous in the twenty-first century, when the New Zealand film industry was acknowledged as having a record of unparalleled achievement.

"RECOMMENDED...darkly humorous."
Chicago Herald-Tribune.

"ACCOMPLISHED...remarkable and restrained assurance."
London Daily Telegraph.

"HONEST, worldly-wise and often very funny."
The Dominion, Wellington.

"ACHIEVEMENT...recognisably authentic."
Auckland Star.

"PICTORIALLY STRONG...dramatically effective."
N.Z.Herald.

"GENUINELY SUBTLE....intelligent and perceptive."
N.Z.Listener.

SKIN DEEP

A film by **Geoff Steven.**
Introducing DERYN COOPER as Sandra.

THE SPA HEALTH CLUB & SAUNA

FOUR

SEX, WHIPS
AND FEATHERS

*T*he first meeting of the interim Film Commission was held in Wellington on November 20, 1977. With $600,000 from the Lottery Board to spend on film-making, and $46,000 for administration, it had to advise the government on how to establish a permanent commission and help develop a sound motion-picture industry. Chair Bill Sheat said he wanted an informal and open relationship with the film industry.

The New Zealand Film Commission Act was passed almost twelve months later. When Alan Highet introduced the legislation, he made a speech which has been quoted ever since: 'We need our own stories and our own heroes. We need to hear our own voices.'

The interim commission's staff of three worked in a dark mezzanine space above the lobby of the Wellington Opera House, a historic building which had been saved from demolition by a campaign led by Sheat himself. Don Blakeney, known to most people as Scrubbs, was the first executive director. Don had been hired for his international financial experience. A chartered accountant, he had worked at the P&O shipping line in London, where he'd learned how tax-shelter deals could help international projects. He had then become an international surfer, before arriving home in time to get the job of caterer on *Sleeping Dogs*. Kerry Robins, later to become a Wellington cinema operator, was his advisory officer. Jackie Aldridge was the first secretary, and Blakeney's dog was the other inhabitant of the office.

'The thing about films in those days was that we were a family,' Don would say later.

On my first visit I patted the dog and told Don that the commission should start marketing the films it was financing. I wanted to be the person who did this. I wanted to create international audiences for New Zealand films in the same way I had created New Zealand audiences for international films. I delivered the same message to Bill Sheat. Then I held my breath and went on with my routine as a television news producer.

The new commission met for the first time on November 13, 1978 in a borrowed office in Auckland. John O'Shea was chosen as deputy chair and the other board members were David Gascoigne; Royce Moodabe of Amalgamated Theatres (the second-largest cinema chain, wholly owned by 20th Century Fox); Merv Corner, a retired banker representing the Lottery Board; and Davina Whitehouse, a British-born actress who had appeared in *Sleeping Dogs* and *Solo*.

The first feature financed by the interim commission (and also by the Arts Council, Montana Wines, a private investor and the director) was released the month parliament passed the act. *Angel Mine* was made for $13,500, plus $26,000 to blow it up from 16-millimetre to 35-millimetre for theatrical screenings. The film, about the fantasy lives of a young suburban couple whose alter egos commit unspeakable deeds while wearing skin-tight black leather outfits, was advertised as New Zealand's 'own erotic fantasy that's far too close to home'. It had been made by 23-year-old David Blyth, who had given up law studies at Auckland University when he decided movies were more interesting than the law.

The censor restricted the film to audiences aged 18 or over, and added a warning that it contained 'punk cult material'. Alan Highet was invited to the premiere, where his parliamentary sensibilities were disconcerted by scenes of sex, nudity, whips and feathers. But he had to accept that the Film Commission was, in his own words, a fully independent statutory authority. Hamish Keith, chair of the Arts Council, wrote in the *Listener* that *Angel Mine* was 'an amazing black comedy cut from the cloth of New Zealand suburban life ... rather than some moving myth about the land'. Patricia Bartlett was less impressed: the Arts Council's grant to the film was an insult to the Queen, and the commission's investment a national scandal.

The second feature film to win support was Geoff Steven's *Skin Deep*, a story about the clash of values when a massage parlour sets up business in a small town. Steven had co-written the film with Roger Horrocks and lawyer Piers Davies. The $180,000 budget was made up of a cautious $70,000 from the commission, a $10,000 distribution guarantee from Amalgamated Theatres, and a mix of private investment, postponed fees, and sponsorship deals. In February 1979, the month the film opened, I achieved my ambition of escaping from television and joining the commission's staff.

Before *Skin Deep*'s local release, producer John Maynard had taken it on a promotional tour through Europe. A screening in Paris for French critics had been arranged with the help of a New Zealand diplomat. Then there had been an American premiere at the Chicago Film Festival, where critic Roger Ebert said the film had 'a nice feel for small-town life and its hypocrisies'. *Variety* called it 'New Zealand's long-awaited breakthrough film' – a quote I exploited to the hilt in Film Commission publicity. Geoff also went to Sydney for a release in the Opera House cinema. He told *The Sydney Morning Herald* the cost of the film was equal to about 30 seconds of *Superman*, or 15 seconds of *Apocalypse Now*.

John Maynard was an Australian who had developed his interest in film after coming to New Zealand at the end of the 1960s to be founding director of the Govett-Brewster Gallery in New Plymouth. His influence extended to the Museum of Modern Art in New York, where *Skin Deep* was selected for opening night at the annual New Directors New Films festival. Every year for the next two decades I would try to persuade the New York decision-makers to choose more New Zealand films. They would select very few.

By the time I joined the commission it had moved from Wellington's Opera House to the second floor of a green art deco building on The Terrace. The staff had increased by two: Russell Campbell handled applications part-time and Lynette Gordon did the same with the accounts. At a staff meeting, Don Blakeney asked everyone to write their own job descriptions. Lynette added 'nurse' to her list of duties. After some debate, the staff agreed to accept collect phone calls.

My official title was head of marketing and information, but the office was anything but formal. We had frequent, unannounced visits from film-makers hoping to persuade the commission to provide money to develop their ideas. Don and I worked separately, but our doors were always open and we talked endlessly.

We both attended board meetings and were encouraged by Bill Sheat to participate in debates. During a break at my first meeting, I learned from Royce Moodabe that the film industry judged the success or failure of films solely on the amount of money collected from ticket sales. Royce had an impressive recall of the dollar earnings of films showing in New Zealand. Until then I had assessed movies by how they were received by critics. When I started a list of the most successful New Zealand films, I followed Royce's example by using box-office figures. Later, to show that there was more to life than money, we added audience numbers as well.

The film industry would always demand judgements, and there were many ways of being judgemental: gross box office, net rentals, how much of the budget had been recouped, audience numbers, critical acclaim, awards. We would learn that films could be important and valuable even if attendances and earnings fell short of hopes and expectations. But film-makers would always be disappointed if their films failed to attract admiring crowds.

One of Don's tasks was to seek private investors, to boost the small amount of production money available from the commission. He wrote two papers for limited circulation, one 'Investment in New Zealand Feature Films', the other a discussion paper on taxation incentives. 'Investment in film production would appear to be a high-risk venture,' he wrote. 'In reality it need not be ... By accident rather than design, the taxation and financial situation in New Zealand is conducive to film production.' He suggested the new commission could make non-recourse loans to private investors. These loans could then be included in the investors' tax returns as part of the money they had invested.

Five of the first twelve feature films in which the commission invested would also have finance from private New Zealand sources. Don's success in this area would become more substantial than anyone expected.

On my first overseas trip I was asked to look for investment money offshore. I found two Americans who said they were interested. One wanted 'a film which would make your wife want to get out and take you to the movies'. The other asked if we could make a film about an oil-tanker disaster. This would be echoed during my first visit to Japan, where distributors said they would pay a million dollars if we could deliver a film about a submarine disaster, preferably starring Kirk Douglas.

The commission, meanwhile, needed a letterhead and logo. Gaylene Preston, who had studied at Canterbury University's Ilam School of Fine Arts, had prepared a concept using a cabbage tree, but I felt we needed an image with a stronger connection with film. I had admired Gordon Walters' painting *Waiata* at the Dowse Gallery, so I wrote to Walters, asking if the commission could use the painting as the cover for its first film catalogue. He said no, but offered to create an entirely new design. I used this first on the front and back covers of the catalogue, which we printed in March 1980. The design attracted international interest and admiration, and in 1981 Walters agreed we could use it as the commission's logo. It has been used ever since. It looks like an unfolding fern leaf which could also be an unwinding spool of film. Art critic Anna Miles wrote in *Art Asia Pacific 23* that Walters' design helped establish a visual language which barely existed before, based on the combination of Maori and modern design.

I established a newsletter and named it *NZFilm*. We created a mailing list of film people and researched the names and addresses of international distributors, starting with the small number I knew. The first issue recorded that the interim commission had given money to 19 feature-film projects, and the new commission was considering 14 more. Another 32 projects had been declined or had lapsed.

Two of the 19 projects were to be produced by John O'Shea. The first was *Sons For The Return Home*, based on Albert Wendt's novel about a love affair between a Samoan man and a New Zealand woman. Wendt, the Pacific's most famous writer, came from Apia in Western Samoa, and had both German and Polynesian ancestry. The film was written and directed by *Landfall*'s director Paul Maunder.

When the film was shot early in 1979, Vincent Ward, a 22-year-old who was about to graduate with a Diploma of Fine Arts (Honours) from Ilam, took the job of art director. Ward was also a movie-maker: his 50-minute drama *A State Of Siege* – based on a novel by internationally renowned New Zealand novelist Janet Frame – had premiered at the Wellington Film Festival the previous year and had won a Gold Hugo at the Chicago Film Festival. The director of the San Francisco Film Festival had called it a work of genius.

A year after *Sons For The Return Home* came the second O'Shea production. *Pictures* was loosely based on the lives of nineteenth-century New Zealand pioneer photographers the Burton Brothers. The cast included Terence Bayler, who had starred in *Broken Barrier* almost 30 years earlier and was now an actor in Britain. Like *Broken Barrier*, the new film explored relationships between European and Maori cultures. Playwright Robert Lord came home from New York to write the script, and director Michael Black contributed original research. 'A scissors-and-paste job, combining Bob's dialogue with Michael's visuals and not really coming to grips with the guts of it,' would be O'Shea's judgement in his memoirs.

From the start of my job at the commission I set an open-door policy. The commission would sell any films made by independent New Zealand film-makers, whether or not it had invested in them. This would continue until the volume of independent production became too great for the commission to handle without increasing its staff, which it was reluctant to do.

One of my first tasks was to meet a group of film-makers who had spent three years organising New Zealand participation at an international television programme market in Cannes. Having already persuaded the commission to take over New Zealand's stand at the market, they handed me a shoebox full of correspondence. Bookings had been made. Two months after joining the commission I flew to the south of France – to the world's largest television market, MIP-TV.

In 1979, MIP was held in the glass-fronted Palais des Festivals on the Croisette, the broad boulevard which runs along the beaches on the edge of the Bay of Cannes. Built 25 years earlier as the permanent home of the Cannes Film Festival, the Palais was also used for international markets – one every month, it seemed – supported by the city of Cannes.

Our small New Zealand stand was at the front, looking out at the Mediterranean. I shared the space with Tom Williamson, representing the National Film Unit, and Dave Gibson and Yvonne Mackay, who had been making short films for the Education Department and were looking for larger audiences with a 10-minute film called *High-Country Children*. Sam Pillsbury, now an independent film-maker with a half-hour film titled *Against The Lights* and based on Witi Ihimaera's story 'Truth of the Matter', was also with us.

Television New Zealand, represented by its programme sales manager David Compton and its controller of programmes Des Monaghan, was in a different part of the market, without a view. Four years later I would succeed in persuading them to join us in a combined New Zealand promotion, but the arrangement wouldn't last long.

South Pacific Television, a subsidiary of the state television company, wasn't visible at all. Its admirable New Zealand productions – including *Children Of Fire Mountain*, *Hunter's Gold* and a series of four feature-length dramas based on stories by Ngaio Marsh – were being marketed and sold by a British television company, which I suspected was not mentioning their New Zealand origins.

I felt strongly that New Zealand product should be sold under the New Zealand brand, and this was the policy the commission agreed to follow. For the next ten years we used the words 'New Zealand' in large black capital letters as the headline at the top of all our advertising.

Not everyone agreed with the idea of the commission selling films. 'I don't think the NZFC should be a sales agent,' producer John Barnett told two visiting Australian journalists. Apart from the comedy *Middle Age Spread* and the telemovie *Iris*, he never allowed the commission to represent his productions, even when it had invested in them. He preferred overseas representatives and insisted on having them. He believed that films should be sold by genre and not by nationality. We would always disagree on this subject.

Eighty of the New Zealand titles at MIP had been made by independent film-makers, an indication of the vigour of those who had campaigned to get state support for film making, and were now the core group hoping to make local features. A lively example was a slapstick series of five six-minute *Percy The Policeman* episodes made five years earlier by Geoff Murphy and the Blerta team at Pacific Films, with Bruno Lawrence as Burglar Bill. In international television terms five short episodes was hardly a series, but we found buyers.

I was particularly attached to a half-hour film called *Killer Whale*, a documentary made by Simon Cotton, an Auckland doctor. Although the film delivered only a few blurry seconds of its subject, the title was enough to persuade television stations to buy it, with so much success that the doctor sent flowers in gratitude for the licence fees we earned for him.

Michael Firth had asked me to represent his skiing documentary *Off The Edge*, which had been nominated for an Academy Award. I sold it to a dozen television broadcasters, including the biggest network in France. Firth's film also became the first New Zealand feature shown by Home Box Office, the wealthy new American pay-television system, which had two million subscribers. As a result, we hosted a visit to New Zealand by HBO boss Frank Biondi and his wife. But the fast-growing network soon focused solely on Hollywood and its early interest in New Zealand films evaporated.

The Cannes Film Festival was to take place a few weeks after the MIP market. I decided not to stay; there hadn't been time to prepare, and I knew little about the festival apart from what I'd seen in the prescient television documentary *Lost In The Garden Of The World*, made five years earlier by Tony Williams and Michael Heath. Heath had interviewed a group of then-unknown directors (including Steven Spielberg, Martin Scorsese, Tobe Hooper and Werner Herzog) and lamented that New Zealanders had nothing cinematic of their own.

My MIP visit to Cannes was, however, a chance to decide how to make New Zealand a part of the next year's festival. The Australian Film Commission, which had been attending Cannes since 1975, had just parted company with a local woman who had been helping organise its participation. Monique Malard was a former Pan Am airline staffer who had moved to the south of France from Paris with her husband, Christian, and their two daughters. I telephoned her. She agreed to help with our inaugural participation. Thus began a friendship of more than 20 years, with the blonde and observant Monique becoming not only the welcoming front-of-house face of New Zealand film at Cannes but also the behind-the-scenes confidante of many New Zealand film-makers and their companions. She guarded her secrets well.

If I had attended the Cannes Film Festival in 1979, I would have seen the premiere of the Australian feature *My Brilliant Career*, with Sam Neill acting in his first leading role outside New Zealand. Instead, at the Wellington Film Festival two months later, I saw *Red Mole On The Road*, the last film he had directed at the National Film Unit.

The Wellington festival also premiered *Sadness Of The Post-Intellectual Art Critic*, a first film by Wellingtonian George Rose. The film's set was built by Tim Bevan, later to become a major international producer with his London company Working Title. Its sound recordist, Don Reynolds, would become a prolific local producer.

I got back to New Zealand in time for the release of the first feature film directed by former film society chair John Reid. *Middle Age Spread*, a comedy about the mid-life crises and marital infidelities of a school teacher, had been adapted from a popular stage play written by former civil servant and teacher Roger Hall, and made with Film Commission money. 'One is tempted to the view that the film improves on the play,' wrote *The Auckland Star* haughtily. Michael Horton had edited the film; he was one of many beginning an independent career after leaving television. More than two decades later he would earn an Academy Award nomination for his work on *The Two Towers*, the second in *The Lord Of The Rings* trilogy.

Middle Age Spread ran for eight weeks in the Wintergarden in the basement of Auckland's grand old Civic Theatre and seven at the Lido in Wellington. It was named as one of the ten best films of the year by *The Dominion's* Catherine de la Roche, who praised the performance of Grant Tilly as the unfaithful teacher as 'superb … he juggles comedy and drama so deftly you can hardly tell t'other from which'. As the first half of the programme we showed *All The Way Up There*, a half-hour documentary by Gaylene Preston and Warrick Attewell about Bruce Burgess, a young paraplegic, being helped by mountaineer Graeme Dingle to climb Mount Ruapehu.

Middle Age Spread was seen by more than 70,000 New Zealanders, and could have been seen by more had it been released on the 35-millimetre commercial gauge, rather than 16-millimetre which limited its local theatrical life. It became a popular seller internationally, and was the first of many New Zealand features I licensed to the BBC. During my first meeting with BBC programmer Alan Howden I said we could offer him an original-cast movie. The play was then having a successful run in London, and Alan at first assumed I was talking about the British cast. He learnt that an original cast, in this case including top New Zealand actors Donna Akersten and Dorothy McKegg, was not necessarily British.

Sons For The Return Home, released through Kerridge Odeon, then the biggest cinema chain in the country, became the third local feature seen by New Zealanders in 1979. The new legislation helped the censor cope with a modest sex scene filmed in a cemetery: he approved the film uncut for exhibition only to persons 16 years of age and over, or to persons under 16 if accompanied by a parent or teacher.

This was my first chance to get involved in a New Zealand release. We persuaded Longman Paul to release a paperback of the novel with the movie poster on the cover. *The Auckland Star*'s weekend edition serialised the book, author Albert Wendt came to New Zealand for a week's tour, and Radio Pacific broadcast a three-hour programme for Auckland's Samoan community. Our publicity noted that the film was the first feature shot in Samoa since *Return To Paradise* starring Gary Cooper in 1952. Moira Walker, who played a supporting role, had begun her acting career in this Hollywood movie.

At the world premiere in Auckland's beautiful St James Theatre, the Queen of Tonga was guest of honour and the screening was followed by a Samoan banquet. Michael Heath, reviewing the film for Wellington's *Evening Post*, wrote that it recalled 'the visual grace that dominates the work of Indian director Satyajit Ray – especially in the Western Samoa scenes which are the best scenes in the film.' In *The Auckland Star*, Nicholas Reid named it as one of the year's ten best films, and the Catholic newspaper *Zealandia* called it 'a film which every New Zealander should see'.

The film had not, however, been without its problems. During post-production there had been a dispute between O'Shea and Maunder over how to edit it, and Don Blakeney had had to decide whom to support. It was a difficult decision because of O'Shea's mana in the industry and his influence in getting the commission established. However Don decided the commission should back the director, rather than the producer. O'Shea was angry. He took his name off the film and the feature was released with no producer credit, although Pacific Films kept its billing as production company.

The only other problem *Sons For The Return Home* struck was, ironically, in Samoa, where its modest frankness was considered shocking. The censors insisted on cutting the love-making scene before allowing the film to be screened.

The first four commission investments had all been for first-time feature-film directors, but Geoff Murphy and Roger Donaldson had proved before the commission was established that they could make successful films. It wasn't hard to guess that they would soon get commission finance.

Murphy was given $5,000 to develop an action comedy called *Meatball*, though when a Canadian film with a similar name was released the title was changed to *Goodbye Pork Pie*. The film was first announced with Ian Mune as director and Murphy as producer, but by the end of 1979 Mune had withdrawn to concentrate on another project and Murphy was directing it himself – on locations from Kaitaia to Invercargill. The film was financed by the commission, with private money from New Zealand United Corporation, three Minis from the New Zealand Motor Corporation, and deferred fees from the film-makers.

The same year Donaldson received $3,000 from the commission to develop *Smash Palace*, although the film didn't go into production for two more years because the commission kept asking for script revisions. 'There was a stage when it got pretty tense ... I guess the resultant anger in Roger may have released some creative juices and helped him achieve the final result ... which as we all know is a great movie,' was how Bill Sheat remembered it.

During the long development period Donaldson was not idle. He directed a privately financed, short family feature for producer John Barnett, and started writing the script of a feature film he called *The World's Fastest Indian*, a true story about a motorcycle enthusiast from the small New Zealand city of Invercargill who, in the 1960s, set a new speed record at Bonneville Salt Flats in the United States – while in his seventies. *The World's Fastest Indian* stayed on Donaldson's wish list for 25 years, while he made eleven movies in Hollywood. It would finally go into production at the end of 2004.

Also privately financed in 1979 was *Squeeze*, a film about an Auckland business-man who has to choose between his fiancée (played by Donna Akersten) and his boyfriend. It was the first feature directed by Richard Turner, who had made four documentaries about New Zealand poets, with finance from the Education Department. He and his producers were unhappy when the commission declined investment. Patricia Bartlett's 'community standards' clause got the blame, and Turner moved to Australia after what he later described as the 'terrible experience'.

When the film was completed I listed it in the commission catalogue. But apart from a small release in the United States it didn't become an international seller, in part because there seemed to be only one release print, which I could never get hold of to screen for potential buyers. *Squeeze*'s editor Jamie Selkirk – another who had escaped from television – would achieve fame two decades later as the Academy Award-winning editor of *The Return Of The King*.

Inevitably there would be frustration and dissatisfaction when the commission said no. But it said yes a lot. By the end of its first year it had put money into films totalling over 14 hours of screen time. Nicholas Reid wrote in *The Auckland Star*: 'We are at the stage where each feature film is an adventure – an experiment in seeing just how much local talents and resources can achieve.' Film student Karl Mutch, in an unpublished thesis, argued that by the end of the '70s the film industry had reached a similar stage to New Zealand poetry in the 1930s or New Zealand fiction in the 1940s, providing increasingly mature and sophisticated glimpses of New Zealand life.

At the beginning of 1980 the administrators of the Japan-New Zealand Cultural Exchange Agreement decided New Zealand culture should be taken to Japan, and I was chosen to deliver it. I flew to Tokyo for a week to apprise Japanese distributors of the existence of New Zealand films – and to learn about the tastes of Japanese cinema-goers, who were enthusiastically viewing the Australian movie *Mad Max*. I met six theatrical distributors and six television sales agents, as well as three television channels and the national organisations which represented producers and distributors. The theatrical distributors, with schedules dominated by Hollywood, asked if New Zealand could provide films with violence, war stories or animals. But New Zealand didn't have any animal films, or *Mad Max* equivalents.

From Tokyo I flew to New York for my first meetings with American theatrical distributors, who would prove to be more open to New Zealand movies than their Japanese counterparts, but not to the first film I offered them. I had sent a print of *Sons For The Return Home* to New York and booked a series of screenings. After the first two I cancelled the others. The film had no appeal to Americans.

In less than 18 months the situation would have changed, and we would be signing the first contracts for New Zealand films to be shown in the United States.

Goodbye
Pork Pie

Written by GEOFF MURPHY and IAN MUNE
Starring TONY BARRY ● KELLY JOHNSON ● CLAIRE OBERMAN
● SHIRLEY GRUAR
Produced by GEOFF MURPHY and NIGEL HUTCHINSON
Directed by GEOFF MURPHY

FIVE

WHICH WAY'S
INVERCARGILL?

BACK HOME, I STARTED PREPARING to return to Cannes for my second MIP market – and, more challengingly, to take feature films to the Cannes Film Festival for New Zealand's first official participation in the world's biggest annual film event.

Monique Malard had found us an apartment to use as an office. She contacted a friend who managed the Ambassades Cinema and made bookings for New Zealand screenings. I persuaded Geoff Murphy to speed up post-production of *Goodbye Pork Pie,* his riotous road movie about two young men trying to avoid the police as they race from Kaitaia to Invercargill in a stolen yellow Mini. It seemed to be the right film to introduce New Zealand to the world's film distributors.

Geoff reluctantly agreed to deliver a first print just in time, but he didn't come to France. He had worked for almost two years on a film whose total budget was only $350,000, and had invested his fee in the film. He chose instead to go to Australia for 'a job I would get paid for', creating special effects for a television mini-series about Ned Kelly.

Goodbye Pork Pie's producer, Nigel Hutchinson, took Murphy's place at Cannes. A Scot born in Aberdeen, Hutchinson had worked as a film publicist in London with actor David Hemmings. He had come to New Zealand five years before as co-founder of Motion Pictures, the country's first company to offer film equipment for hire, breaking the Film Unit's monopoly.

I also took a print of *Sons For The Return Home,* and expecting that the world's press would want to interview a Samoan actor I invited Uelese Petaia, who played the lead role of Sione. Kerry Robins was the fourth in the New Zealand delegation. He proved to be a hard-working and practical colleague, and an enthusiastic and constructive troubleshooter.

To keep costs down, our office apartment was three blocks back from the beachfront Croisette where, we belatedly discovered, most of the buyers and sellers were to be found. There were other problems with our frugal venue. The entrance was on one street while the balcony faced a different one. We missed out on buyers who couldn't cope with looking up at us without being able to find our front door.

Another bad idea was our decision to sleep in the apartment's two small bedrooms. During MIP, most New Zealanders stayed in a crumbling old hotel called the Suisse, which we grew to view affectionately in spite of its ancient wardrobes with doors which wouldn't close, and an erratic kitchen where an occasional crisis meant the cancellation of breakfast. But for our first film festival we decided to save money by living and working in one space. It wasn't a success. Nor was our decision to do our washing in the apartment. Our first load of clothes emerged from the small French washing machine a universal shade of orange.

But nothing really worried us as we handed out our brochures and screening schedules, put them in piles in the main hotels, delivered them (in personally addressed envelopes with hand-written letters) to hotel rooms, and copied other penurious film-sellers by placing them under windscreen wipers on parked cars, where we hoped they would create no-cost visibility on the streets. Most of all we talked. Every time we met new people we told them that New Zealand films had arrived. There were endless numbers of people to hear our message. It seemed to be the first time any of them had heard of New Zealand.

Monique organised tickets to the festival's formal opening night. We shared the strange experience of putting on dinner jackets to go to the movies, and then sitting in a gigantic cinema with 2,000 equally formally dressed film-goers. This was the style required for attendance at every evening screening of the films officially selected by the world's pre-eminent film festival. We wanted New Zealand to be a part of it.

It would take four years to get a New Zealand film selected for the competition. But in the market we got instant results. Audiences came to our screenings in search of something new. Among them were distributors who enjoyed *Goodbye Pork Pie* and wanted to buy it. They wanted to sign contracts. I didn't have any knowledge of this subject, but I had met a Paris-based sales agent, Jeannine Seawell, who had made her name introducing Australian films to Europe. I asked for a favour: would she show me one of her contracts and explain it to me? She was willing to help, and I was grateful and relieved.

Jeannine, who would remain a generous friend of New Zealand cinema through-out her life, gave me a copy of a standard international film sales contract. There were many items to be negotiated, including percentage shares of revenue (which varied between theatrical, video and television income), what deductions would be allowed to cover expenses and costs, and what obligations had to be accepted by the distributors, who always promised to use their best efforts to promote and release each film. Best of all was the need to negotiate an advance, and how it would be paid.

I used Jeannine's contract to write the first sale for *Goodbye Pork Pie* – to a buyer from the Middle East who didn't speak English. We negotiated through his son, who translated. The experience caused Nigel and me some nervousness, especially when the contract was signed and the buyer handed over a cheque for the deposit, which we had to carry home. Did he know what he had signed? Would his cheque bounce?

The stress of making our first deals was more than we had anticipated. In the weeks before the festival, Nigel had been working long hours in Wellington to complete the sound mix and oversee delivery of the first print, which he had then packed in his suitcase just before he began the 40-hour trip to France. At Cannes, like the rest of us, he continued to work 18-hour days, and after we signed the first deal he seemed to need medical attention. Monique found a hospital with English-speaking doctors who sent an ambulance to our office door. They diagnosed exhaustion, and put Nigel to bed for the last two days of the festival – a compulsory 48 hours of sleep with the help of sleeping pills before he flew to London.

We ended Cannes with six signed contracts: they were for 20 countries which would release *Goodbye Pork Pie*. The film had created a 'buzz' in the marketplace, something everyone wanted for their films but few achieved. And the cheque from the Middle East didn't bounce.

Sons For The Return Home wasn't so popular. We couldn't find any buyers for it, and I couldn't find any journalists to interview its leading actor. Uelese didn't enjoy the crowds or the lack of activity. Some days before the end of the event he left the office without saying goodbye and flew back to the South Pacific.

On the morning after the festival ended Kerry and I were the last New Zealanders left in town. We looked out towards the beautiful Bay of Cannes. It was a perfect day. The sun shone. The Mediterranean was seductively blue. The deckchairs were laid out on the perfectly groomed sandy beaches where the waiters were ready to serve drinks and meals. But we wanted to leave. We packed our bags, locked the apartment and took a taxi to Nice.

At the airport we bought a copy of the London *Observer*, in which critic Philip French wrote that New Zealand had conducted 'a vigorously publicised intervention in the festival marketplace, with a brisk, inventive road movie and a sombre account of a blighted affair … Both treat familiar themes in a fresh setting and ought to be seen here.'

I flew home through London, Paris, New York and Los Angeles, setting a pattern of travel which would continue for two decades. Twice a year on my way to the big European markets, and twice a year on my way home, I would stop in major cities to get to know distributors and tell them about New Zealand films and why they should consider releasing them. The film business was based on such personal contacts. It was easy to fill my diary for meetings all day and every evening as well.

First I talked to the distributors from whom I had been buying films. But they preferred directors and titles from countries whose film industries were better known and longer established. New Zealand's first theatrical distributors would come from a different group, often beginners like us. With their help we would begin to establish New Zealand as a new source of innovative, entertaining films.

I also sought out festival directors. One of the first to show an interest was the director of the London Film Festival, an American named Ken Wlaschin. He invited *Goodbye Pork Pie* to his festival in November. By that time Geoff Murphy had come home from Australia with his Ned Kelly earnings and had fine-tuned his film's visuals and soundtracks with editor Michael Horton. He flew to London with the re-edited print in his baggage while we began learning how to deliver prints and publicity materials to international distributors.

London was not Geoff's first film festival. The previous year we had arranged for him to take *Wildman* to the New Delhi film festival, where he had explained to Indian journalists that his production house, the Acme Sausage Company, did not make sausages. A *Times of India* critic had written that the photography captured the untamed splendour of New Zealand scenery but the characters were 'people one would never miss having met'.

Within a year *Goodbye Pork Pie* had been sold to more than 30 countries, with Jeannine Seawell handling further sales in Europe. The film was within sight of becoming profitable for its private investors and for the commission.

The London Film Festival screening helped get a United Kingdom release, and it also attracted interest from Buckingham Palace. We learned that Prince Charles studied the festival programme and occasionally requested films for a private screening. He sent his *Pork Pie* request to New Zealand House, and we arranged for a print to be delivered to the palace so he could have his own Christmas holiday screening. We never heard whether he enjoyed it.

We didn't hear much from the British distributor either. The film's modest earnings always seemed to be swallowed up by substantial distribution expenses – something we learned was not uncommon in the film business. Years later, when the contract was about to end, we received our first and only cheque. We framed it and hung it on the office wall. It wasn't enough to cover the bank charges.

Geoff Murphy persuaded the commission to let him be the New Zealand distributor for *Goodbye Pork Pie*. He decided that February 1981, nine months after its Cannes launch, would be the best time for the film to be released in New Zealand. It premiered at Wellington's Majestic Theatre to an enthusiastic audience of 2,000.

Geoff and his family worked mightily to create nationwide awareness of their film. They filled a Mini with cans of Fresh Up fruit juice, and offered the car as a prize to anyone who could guess the correct number. Car dealers all over the country displayed Minis full of Fresh Up cans. Later more than a hundred Minis drove up Auckland's Queen Street to advertise the film.

It went on to set New Zealand box-office records, with more than 600,000 people buying tickets. Income exceeded $1.2 million. Twenty years later the film would still be on the list of the ten highest-grossing local releases of New Zealand films.

At the commission we issued press releases announcing each international sale. We highlighted a *Los Angeles Times* review by Kevin Thomas, an American critic who had come to New Zealand as one of the commission's first guests, describing the film as 'in its zesty way, a work of integrity, anti-materialistic in its sentiments and capturing both the humour and frustration in New Zealand life'. Thomas had also written that New Zealand films were an unknown quantity in the United States 'and, I daresay, in the rest of the world'. We didn't highlight this comment.

I sent *Goodbye Pork Pie* to many festivals, including New Delhi where *The Statesman*'s reviewer deadpanned that it 'gives us many glimpses of New Zealand's beautiful landscape and mountainscape'. The same landscape stopped the film getting a sale to Japan: there were so few cities and towns that distributors couldn't believe the film's anti-heroes were travelling a long distance. They said Japanese film-goers would expect to see many buildings on such a long journey.

After the experience of Cannes, I arrived home worrying about how to find some international interest in *Sons For The Return Home*. We achieved this with the help of a Czech-born New Zealander called Michael Havas, who had introduced himself to us at Cannes. Havas had grown up in Auckland and graduated from Auckland University before deciding to visit his parents' homeland, where he had been living and studying for nine years and seemed to know everyone in the film community. He had graduated with honours from the Film Academy in Prague after completing a paper on New Zealand cinema.

Michael used his contacts to get the film selected for the Karlovy Vary Film

Festival, where Uelese Petaia won the award for best actor, tied with Al Pacino in Norman Jewison's *And Justice For All*. Paul Maunder collected the award, and the London *Daily Telegraph*'s critic wrote, 'It was a relief to come across such a gentle, humane and serious-minded film.' Later, with Michael's help, we sold the film for theatrical release in Czechoslovakia and for telecasting in East Germany.

While Geoff Murphy had been forging ahead with his story of two young men in a small yellow car tearing around New Zealand trying to evade the police, Wellington producer John Barnett had completed a feature which also involved the law. *Beyond Reasonable Doubt* was based on the best-selling book by British author David Yallop about the police investigation into what had become known as the Crewe murders. Jeanette and Harvey Crewe, a young farming couple, had been shot dead in their living room while their baby slept nearby. Later someone had fed the baby. The murders had transfixed the country, with people arguing around dinner tables as to who had had the means and the motive.

John Laing, who had started his career in 1970 at the National Film Unit, came home to direct the feature after three years editing at the BBC and five years at Canada's National Film Board. For the Film Commission it seemed a landmark that a New Zealand film-maker had been enabled to come home to work.

The commission's investment in *Beyond Reasonable Doubt* had been boosted by money from three private sources: investment company Brierleys, property developer Bob Jones and merchant bankers Fay Richwhite. The names of these three investors appeared on screen before the title of the movie, in the position conventionally occupied by the MGM lion or the Paramount mountain. Local audiences weren't accustomed to seeing familiar local names on the screen. At one of the previews, the credits brought laughter from the audience. It was the same embarrassed laughter I'd heard as a child, when a local politician flashed by during a newsreel.

David Hemmings, who had been the fashionably alienated young photographer in Antonioni's *Blow-Up* 15 years earlier, had come from London to play Inspector Bruce Hutton. Arthur Allan Thomas was played by Australian actor John Hargreaves. While the movie was in post-production, Thomas was pardoned and released after nine years in jail. It appeared that the police had planted a cartridge case in order to ensure a conviction. The change in circumstances meant the script had to have a new ending; the commission asked for lawyers to check and ensure that every word and event was authentic.

Beyond Reasonable Doubt opened with a five-week run at the Civic in Auckland. Although its total national attendance of 150,000 didn't equal the box office achieved by *Sleeping Dogs*, it broke some of the weekly records. All these records would be beaten again when *Goodbye Pork Pie* was released six months later.

David Hemmings discovered new opportunities during his New Zealand stay, and in 1981 he decided to return as a film-maker. With the help of tax-sheltered investment from Fay Richwhite, he spent five months in Queenstown directing an action-comedy called *Race For The Yankee Zephyr*. Everett de Roche's script had originally been set in Australia, but Australia didn't offer the chance for tax deals like those then being developed in New Zealand.

Co-produced by John Barnett and Australian Antony Ginnane, the film starred three Americans and an Englishman alongside New Zealander actors Bruno Lawrence and Grant Tilly. John Laing was employed as editor. Hemmings' British company Hemdale (where Nigel Hutchinson had worked before coming to New Zealand) acquired international sales rights.

Tax deals were to become a big issue, but in 1980 I was barely aware of them as I focused on the need for New Zealand to participate in a second market. At Cannes I'd learned that everyone would be going to Milan six months later for Mifed, an international film and multimedia market, and I thought New Zealand should be there too. The commission agreed, and at the end of the year I again flew to Europe, this time to Italy.

Just before I left, John Maynard introduced me to Dorothee Pinfold, an outgoing New Zealander who had come home for a visit after working in London where she had handled international distribution of *The Muppet Show*. We needed her specialised knowledge about delivering materials all over the world, especially as *Goodbye Pork Pie* responsibilities were starting to overload the staff at Nigel Hutchinson's production company. The commission hired her, and one of her first tasks was to take over the *Pork Pie* files. I welcomed her to the office and flew to Milan.

Because I had registered late, the only space available at the Mifed market was a small table in a corridor at the top of some stairs. The first week was for television sales and the place was almost empty: sales of independent television productions had moved to the MIP event in Cannes. The second week, though, was for feature films. Suddenly the huge Fiera building, designed during the time of Mussolini, was crowded.

I started to discover that the film industry was full of people with backgrounds similar to mine. They had entered it after developing a passion for movies at a cultural level, and were working to create new audiences for films from outside the Hollywood system.

Alan Howden of the BBC came from the British film society movement. Chicago-born Carole Myer was pioneering sales of features made by the production board of the British Film Institute, and overcoming the resistance of distributors who considered them too difficult. Dutch-born Eliane Stutterheim was a partner in a Swiss distribution company whose programming was similar to that of our film societies. Two years later she would win a Cannes Palme d'Or for her production of Kurdish director Yilmaz Güney's autobiographical film *Yol* (*The Road*), a harsh story of prisoners in Turkish jails. Güney had written the screenplay in his jail cell; the film had been directed according to his instructions before he escaped and fled to France.

New Yorker Donna Gigliotti, who had started her career as assistant to Martin Scorsese, was buying films for a newly established 'classics' division of an American studio – 'classics' in those days being an all-embracing word for foreign films not made by one of the major studios. Bostonian Carol Greene, who had studied film at Columbia University under influential New York critic Andrew Sarris, was buying films for American publishing company Macmillan, which was also distributing movies. Soon she would join one of the first pay-television networks. Her friend Sara Risher was involved with a small company named New Line, which was releasing low-budget horror films.

Ben Barenholtz was an independent New York distributor whose successes with non-American films in the US market had encouraged the studios to broaden their interests. Canadian distributor Linda Beath was a lover of French cinema who would soon open a company in the United States with Iranian director Bahman Farmanara (whose film *Prince Ehtejab* I had so admired) as her business partner. Tom Garvin was a young Los Angeles lawyer starting a career working with independent film-makers. He eventually persuaded me that legal advice was necessary when negotiating with the biggest American companies. All of these people would, over the next few years, become involved with New Zealand films.

Jeannine Seawell introduced me to Guglielmo Biraghi, who was directing the Taormina Film Festival in Sicily. A distinguished Italian journalist and playwright, Biraghi wanted to support new film-making countries. He would be the first to show our films in Italy.

Another of Jeannine's friends was Klaus Hellwig from Germany. Hellwig's contacts inside German television led to the first sales of New Zealand feature films to German-speaking territories. I tried to bypass him and sell films direct to the television buyers but the German system was immovable: outsiders weren't allowed in. The buyers were always willing to join us for meals and receptions, and they invited us to their birthday parties which often seemed to coincide with markets and festivals, but they kept avoiding me when I wanted to talk business. Our first three years' sales to Germany were all handled by Hellwig as middle-man; he made several visits to New Zealand to keep ahead of his competitors by being the first German to learn about films in production.

Then there were the many independent cinema distributors from all over Europe. They all had an enthusiasm for films from emerging industries and they knew how to reach the growing number of European cinema-goers who, like Wellington Film Festival audiences, were looking for an alternative to Hollywood. But in another way the European countries couldn't have been more different from New Zealand: their public television systems were willing to support the theatrical distributors by sharing the cost of the advances which had to be paid when films were contracted for release. New Zealand television executives would never have considered such an idea.

Soon I had made a group of close friends who were all in the business of selling or buying films. We would organise a dinner at every market. There were no secrets. If someone found a new potential buyer, all the sellers would all get the details. If someone saw a great new film, all the buyers would be told. If personal relationships changed, as they often did after working hours, we would all be interested too.

American independent production hadn't yet taken off, but the American studios were closely watching the emergence of new film industries. Despite initially opposing the formation of the New Zealand Film Commission, 20th Century Fox chairman Denis Stanfill set up a Fox fellowship for young New Zealand producers. It was first awarded (from 43 applicants) to Timothy White, who had produced Vincent Ward's *A State Of Siege*. I visited White in his temporary office on the Fox back lot, and could see how it was easy to be impressed by the dominance, attraction and wealth of old Hollywood. Fox people told me White was reluctant to leave Los Angeles when the six-month fellowship was over. 'We had to push him on to the plane,' they claimed.

When I started work at the commission, I expected to continue running the Wellington Film Festival. By the end of my first year, however, I had realised I couldn't do both jobs. The commission was all-engrossing, much more so than my final years of producing network news for television.

Dunedin-born Bill Gosden, a student film critic during his years at Otago University, had successfully campaigned for the job of administrator of the film festival and the film societies. He was well-positioned to become the festival's first full-time professional director in 1981. Within a few years, he had also brought the older, lacklustre Auckland Film Festival under his control, setting the foundation for a national chain of festivals.

In the same year Jonathan Dennis became the first director – indeed the first staff member – of the New Zealand Film Archive. Jonathan had come home after spending two years travelling in Europe, learning about film preservation. He had trained at the great European film archives, including the Cinémathèque Française where he had become a friend of Mary Meerson, the companion of its legendary founder Henri Langlois. He had made influential friends at all the archives, and while an unpaid intern in London at the National Film Archive, supporting himself by working as an usher at the National Film Theatre, he had also bonded with many British Film Institute staff. At my first Cannes Film Festival he had taken me to a reception given by the institute and introduced me to Carole Myer.

Jonathan had sent me a trail of postcards from Europe asking, 'Where is the New Zealand Film Archive?' It was a question I didn't understand at first, but as the postcards and the question kept on coming I persuaded a group of film and television people to meet and discuss how such an archive could be created. The Film Unit, the Broadcasting Corporation, the National Archive and the film societies all sent a representative, as did the Education Department's National Film Library, which was storing Walter Harris's priceless collection of early films, including much of the work of Rudall Hayward. The idea won favour: the Film Commission agreed to create a trust, which in turn set up the archive.

In time, the legend would spring up that Jonathan had founded the Film Archive single-handed. In a sense it was true. Certainly his tenacious advocacy and specialised knowledge had made the idea irresistible. 'It might be fairer to say that others had to found the archive because Jonathan had equipped himself to run it,' Bill Gosden would later comment.

Jonathan's parents and friends helped paint and decorate the interior of an old building near the Wellington Town Hall, which was rented as the Film Archive's first temporary home. The most urgent task was to raise money to preserve nearly a million metres of irreplaceable nitrate film, most of which was decomposing and needed copying if it were to be saved. When Jonathan found *The Devil's Pit*, a previously lost silent film from the 1920s, he also discovered that its star was still alive and living in Rotorua. Witarina Harris, then in her eighties, became kaumatua, or respected elder, of the new Film Archive, travelling with Jonathan all over the world to introduce archival screenings.

In April 1981 I headed back to France for my third MIP-TV, with Dorothee Pinfold, Kerry Robins and 50 short films made by independents. This time David Compton was handling sales and marketing for Gibson Films, who were aiming at the international television market with their 50-minute *Monsters' Christmas*.

There was a gap of two weeks between MIP and our second Cannes Film Festival. In this space we scheduled New Zealand's first film retrospective, an idea proposed by Ken Wlaschin when he had selected *Goodbye Pork Pie* for the London Film Festival the previous November.

The New Zealand industry we were promoting was less than three years old, but Wlaschin managed to assemble a ten-day season at the National Film Theatre in London by including 1970s features such as *Landfall* and *Test Pictures*, *The God Boy* – a telemovie directed by Murray Reece and written by Ian Mune from a novel by Ian Cross – and a seven-hour screening of all the episodes of Tony Isaac's ambitious and spectacular historical television series *The Governor*, which had never been seen in Britain. Vincent Ward's *A State Of Siege* and his second 50-minute film *In Spring One Plants Alone* were also on the programme, along with early films by Geoff Murphy, Roger Donaldson, Geoff Steven, David Blyth, Barry Barclay and Sam Pillsbury.

The Observer's Philip French wrote that the retrospective showed 'New Zealand … is encouraging its film-makers a good deal more coherently than we are. And with good results.'

The opening night film was *Beyond Reasonable Doubt*, followed by a reception for 400 people in the ballroom at New Zealand House, hosted by the New Zealand high commissioner. Les Gandar, a Manawatu sheep farmer (and former chancellor of Massey University), had been a colleague of my father's in Keith Holyoake's National government. I asked him to help organise a royal premiere for the British theatrical release of the film. It seemed a reasonable request. But Gandar refused. There was, he said, no way he would allow the royal family to be associated with the film's portrayal of the New Zealand police force. I told him every word of dialogue was true and had been vetted by lawyers, but he wouldn't be moved. The film never got a royal premiere.

Unlike Gandar, British critics didn't take offence. 'Especially interesting for its picture of New Zealand rural life,' veteran critic Dilys Powell wrote in *Punch*. It was probably the first New Zealand film she had ever seen.

SMASH PALACE

"SHE'S MY DAUGHTER AND I'LL DO WHAT I LIKE!"

AARDVARK FILMS PRESENTS
A Film by ROGER DONALDSON

BRUNO LAWRENCE   ANNA JEMISON   KEITH ABERDEIN   GREER ROBSON
DES KELLY   Associate Producer LARRY PARR   Music by SHARON O'NEILL
Written, Produced and Directed by ROGER DONALDSON

SIX

THE SMASH PALACE
SPLASH

OR NEW ZEALAND'S SECOND Cannes Film Festival, we moved the office
out of its back street and into an apartment on the Croisette. On the first
morning we looked out of our eighth-floor office windows, and across the
street we saw a restaurant on the beach, with rows of yellow umbrellas above tables
and deck-chairs, not far from high-tide mark. Although there were more than a dozen
beach restaurants competing for diners, L'Ondine would become my choice for 20
years when I took buyers to lunch. I recommended the marinated raw fish salad, and
marvelled at how so many meals could be impeccably served from a tiny windowless
kitchen that seemed to be under the road.

There were twice as many New Zealanders at Cannes in 1981 as in our first year,
with Film Commission chair Bill Sheat making the first of six visits, and chief executive
Don Blakeney what would be his first and last. We booked market screenings of
Beyond Reasonable Doubt, *Smash Palace* and *Pictures* every afternoon at two o'clock.
This was the start of the commission's generic promotion of New Zealand as a new
(and, we implied, exciting) source of films. It was effective – and frugal. The total cost
of Cannes that year, including travel and advertising, was NZ$31,000.

We met London publicist Soren Fischer, who knew Judy Garland and had
looked after Hollywood stars, including Bette Davis and Gregory Peck, when they
visited London for lectures at the National Film Theatre; he also worked for the
Berlin Film Festival and seemed to know everybody in the European film world.
Soren agreed to help us, for a tiny fee. When he explained the need for messengers
we hired Yves Millet, who became our second local helper, joining us every year
and bringing his scooter, on which many New Zealanders were given speedy rides
through traffic jams to appointments they would otherwise have missed.

Soren also advised us on the mysteries of organising New Zealand's first Cannes
reception. We invited Favre Le Bret, the president of the festival, as our guest of honour.
We hoped this would encourage the festival to consider inviting a New Zealand film
into one of its official sections.

Five producers were with us, one of whom was John Barnett. I was taken by surprise when John spoke at a meeting of producers in our office, surrounded by images of the New Zealand films we were marketing. His subject: his opposition to the commission selling films. His criticism continued later in the year. I wrote to him denying his claim that producers were required to hand over their films to the commission. In reality it was always a collaborative and flexible relationship, just as he had experienced with *Beyond Reasonable Doubt*.

At the start of the festival John hadn't decided who would sell his film. He accepted my offer to help him find distributors, and with his approval I negotiated contracts with Klaus Hellwig for a German theatrical release and with John Hogarth of Enterprise Pictures for a British release. These deals didn't change his opposition to the commission's sales activities: he handed over unsold territories to J&M, a sales agent based in London.

Roger Donaldson, who was at Cannes with his associate producer Larry Parr, was keen for *Smash Palace* to be represented outside North America by the Producers Sales Organisation, a new company run by a former Hollywood actor named Mark Damon. PSO was selling films by Martin Scorsese and Franco Zeffirelli, as well as the German thriller *Das Boot*, which was set to become an international hit. Roger wanted to be a part of this success.

Only Damon could decide what films his company would represent. We tried to get him to see *Smash Palace*. He wasn't available until the day after the market ended, so we booked a screening room for that day. He turned up with several staff and family members, viewed the film – in which Bruno Lawrence gave a bravura performance as a man who kidnaps his daughter after the break-up of his marriage – and left without a word. Six months later Damon agreed to represent the film. PSO proved effective, selling *Smash Palace* to 50 countries in the first year. When the company went out of existence six years later, *Smash Palace* came back into the commission's catalogue.

After our Cannes screenings we received four distribution offers from Americans, and Roger agreed we would handle this sale ourselves. It was a great time for all of us. For Roger, it was the start of international recognition. For me, it was my first chance to negotiate a sale of a New Zealand film for release in the United States. Our industry had been set up to give New Zealand audiences something of their own, in place of American movies. Now American audiences would be seeing New Zealand films.

The four offers were an excuse to make an extra trip to Los Angeles and New York. I scheduled our flights for the end of July, making sure I wouldn't miss any part of the Wellington Film Festival. Roger and Larry flew first class, and on the overnight flight from Los Angeles to New York they met Samuel Z. Arkoff, a Hollywood producer with 40 years' experience in the movie business. He spent the night telling them stories of his career.

Following thrifty commission policy, I was sitting at the back of the plane, from where I occasionally leaned out of my seat to try to guess what I was missing up front. After that experience, the commission was encouraged by board member Royce Moodabe to approve business-class travel for staff making long flights. In future if there was anything to be learned at the front of the plane, I wouldn't miss out. I would be more comfortable too.

In New York and Los Angeles we spent our days negotiating with the four distributors who had made offers to release *Smash Palace*. After a week we decided on a deal with Tom Coleman, who had recently established Atlantic Releasing and had licensed two new Australian films, Peter Weir's *Picnic At Hanging Rock* and Bruce Beresford's *The Getting Of Wisdom*, from Jeannine Seawell.

Smash Palace became the first New Zealand feature film to be sold to the United States, but it would not be the first to get an American theatrical release. That honour went to Roger Donaldson's first feature, *Sleeping Dogs*. I had persuaded Roger to take back the sales rights from a European agent, who had buried the film in a big international catalogue where it was attracting so little interest that Roger was worrying it had no future outside New Zealand. It was a perfect title to re-launch under the New Zealand brand. As soon as I listed it in the commission's New Zealand catalogue, it started to attract buyers.

Within a month of finalising the *Smash Palace* American deal, I licensed *Sleeping Dogs* to Satori, a new and ambitious New York distribution company with offices in the penthouse of a 1930s art deco building on 42nd Street. Its amiable owners Ernie Sauer and Gary Connor had been successful with exploitation films. Now they were aiming to reach mainstream audiences, and willing to take a gamble on films from the South Pacific. For their first Australian release they hired someone to wear a kangaroo suit in the street outside the cinema. To publicise *Sleeping Dogs* they mailed kiwi pins to journalists, and then boxes containing a kiwi fruit.

Satori opened *Sleeping Dogs* in February 1982, just as Kiri Te Kanawa was becoming a star at the Metropolitan Opera in *Cosi Fan Tutti*. The New Zealand consul-general gave a reception for the film and served kiwi fruit and New Zealand wine. Ian Mune flew to New York to help promote the launch. *The Village Voice* noted that *Sleeping Dogs* was the first New Zealand feature ever released in the United States and 'a precociously accomplished' debut. *The New York Times* praised Donaldson's 'sharp, suspenseful direction'.

Smash Palace got even better reviews when its American release began two months later in the Paris, a leading New York art-house cinema on West 58th Street near Central Park, after a premiere in the New Directors New Films festival. Powerful *New York Times* critic Vincent Canby named it as one of the ten best films of the year and hailed Roger as 'a film-maker of potentially worldwide importance, a man of original visions with the technical facility to realise them'.

In *The New Yorker*, famously tough movie critic Pauline Kael appeared incredulous. *Smash Palace* was, she wrote, 'an amazingly accomplished movie to have come out of New Zealand … a remarkable piece of work'.

We particularly liked the *Village Voice* review, in which Carrie Rickey wrote that *Smash Palace* was better than two similar Hollywood productions, Alan Parker's *Shoot The Moon* and Robert Benton's *Kramer vs. Kramer*.

Roger was soon getting job offers from Hollywood, and flying back and forth between Auckland and Los Angeles for negotiations. *The Bounty*, which United Artists had considered shooting in New Zealand a few years earlier with *Dr Zhivago*'s David Lean as director (and with tax concessions over-eagerly offered by the government for overseas technicians and actors), became his first Hollywood movie, with Mel Gibson and Anthony Hopkins as the stars. Roger shot sequences off the coast of New Zealand and used the Whangarei-built ship which had been commissioned for David Lean.

Within a few years, Roger Donaldson had a home and a career in Los Angeles, where he became the first of the new breed of New Zealand film-makers to earn a Hollywood reputation.

Goodbye Pork Pie, *Beyond Reasonable Doubt* and *Skin Deep* were the next three New Zealand features released in United States cinemas. Other titles from those first two years never got an American sale, although we ensured they played in film festivals there. *Pictures* was one that didn't sell to the States, but with help from Jeannine Seawell it did sell to Germany – and the United Kingdom, where it was released by the pioneering art-house company Cinegate and named by *The Daily Telegraph* as one of the year's ten best films.

The Observer's Philip French noted: 'New Zealanders have learnt from the errors of the Australian new wave and wisely restrict themselves to movies with modest budgets and indigenous subjects, made by local talent.'

For many European audiences, *Goodbye Pork Pie* was the first New Zealand film they'd seen. In Spain it was titled *!Vaya Movida¡* and advertised as 'la excitante historia de dos hombres muy diferentes y de su increíble viaje en un Mini Amarillo robado'. For publicity our Spanish distributor drove a yellow Mini through Madrid at rush hour.

The international marketplace wasn't always so friendly. When we took John Reid's second feature, *Carry Me Back*, to Cannes in 1982 we had the depressing experience of finding no one who wanted to buy this comedy of rural life. Things improved six months later at Mifed. We screened the film for Larry and Bonnie Sugar, who represented the American company Lorimar. They loved it and paid a six-figure advance for worldwide rights.

From the beginning the Film Commission, so its money could be spread as widely as possible, encouraged producers to seek investment from private sources. It wasn't long before financiers found a loophole in taxation legislation which they could use to their advantage if they invested in films. The news spread fast. As New Zealand films began their first overseas releases, international film projects started turning up in search of New Zealand investors wanting to take advantage of tax shelters.

This was the start of four years when New Zealand investors enthusiastically discovered they could enter into arrangements to pay less tax if, usually with the help of limited-recourse loans, they invested in films. But many film people were uneasy: would productions from offshore destroy their fragile local industry? The same uneasiness would re-emerge 20 years later with the occasional arrival of huge-budget Hollywood productions such as *The Last Samurai* and *The Lion, The Witch And The Wardrobe*. Such productions were able to pay crew members at a much higher rate than New Zealand productions could afford, creating expectations which couldn't be met by local producers seeking to hire the same skilled people. The result would be impossibly unrealistic pressures on local budgets.

One of the first films to use the tax loophole was *Shadowland*, produced in 1981 by John Barnett and Antony Ginnane and directed by American Michael Laughlin. This murder mystery had David Hemmings as one of three executive producers and was sold by his London company, Hemdale. The underwriters were Auckland merchant bankers Fay Richwhite. Auckland locations such as Remuera, Epsom, One Tree Hill and Avondale College were disguised to look like the United States Midwest. *The New York Times* was impressed: 'This all-American tale happens to have been filmed in New Zealand, but there's nothing about the look of the movie to suggest that Ohio State isn't just around the corner.' The film was renamed *Dead Kids*, and then, for its United States release, *Strange Behaviour*.

An American project, also produced by Barnett and Ginnane, brought Ryan O'Neal's teenage daughter Tatum O'Neal and Broadway star Shirley Knight to New Zealand to shoot a feature titled *Prisoners*. Bruno Lawrence and John Bach were in the cast, as was David Hemmings, again one of three executive producers. Twentieth Century Fox was announced as the worldwide distributor and an elaborate brochure was printed. But the film was never released.

Another overseas project that created media excitement was *Eve And That Damned Apple*, a film created for star Bo Derek and to be produced by her husband, former actor John Derek. The photogenic couple came to Wellington less than three years after Bo Derek had starred opposite Dudley Moore in the 1979 hit comedy *10*, directed by Blake Edwards. They asked the commission to introduce them to politicians so they could discuss the possibility of importing a snake, something forbidden in law. I got the job. There was no difficulty setting appointments.

After our first parliamentary meeting, I walked with Bo Derek along Lambton Quay. She was happy she had charmed the members of parliament into agreeing to modify their ban on snakes, but seemed uncomfortable that no one in the street was recognising her. She didn't relax until I bought a copy of *The Evening Post*, where her photo was on the front page.

The snake loophole never had to be used, as the film was never made.

One of the finest offshore features financed by tax-sheltered New Zealand investment was *Merry Christmas, Mr Lawrence*, made by the distinguished Japanese director Nagisa Oshima. Oshima had made 21 movies, several of which I had bought for film society screenings. *Merry Christmas, Mr Lawrence*, his first feature outside Japan, was shot in Auckland and Rarotonga, with Englishman Jeremy Thomas as producer and Larry Parr, whose former Broadbank colleagues helped raise the finance, as an associate producer.

On the day the prospectus was issued we sent a messenger a few blocks along The Terrace from the Film Commission office to collect a copy. By the time it arrived the film had been over-subscribed. I knew several people who were investors. They seemed very pleased with the deal.

The film, starring David Bowie, Tom Conti and Jack Thompson as Japanese prisoners of war, was selected for competition at Cannes in 1983. Lee Tamahori, who wouldn't direct his first feature until *Once Were Warriors* ten years later, was first assistant director.

Auckland producers Rob Whitehouse and Lloyd Phillips also found private investors for their futuristic action-adventure *Battletruck*. Englishman Harley Cokliss directed the film in the grand scenery of Central Otago. Bruno Lawrence, John Bach, Diana Rowan and Kelly Johnson were in the cast, along with five Americans. The producers pre-sold United States rights to the influential B-grade producer Roger Corman, who released it through his New World company, changing the name several times in an effort to boost the American box office. It was sold to the rest of the world by a company which had just had an immense international success with an American action movie called *Cannonball Run*. The company put *Battletruck* into a package with *Cannonball Two* and it sold everywhere. It became the first New Zealand production to be released in Japan, with 120 prints.

Within a year Whitehouse and Phillips had a deal with Paramount Pictures and more New Zealand investors for *Savage Islands*, a pirate movie starring Tommy Lee Jones in a cast which included New Zealanders Grant Tilly, Kate Harcourt, Bruce Allpress and Prince Tui Teka. They hired British director Ferdinand Fairfax to shoot the film in the Bay of Islands, Auckland, Rotorua and Fiji. For a 1,100-print American release, Paramount changed the name to *Nate And Hayes* but the critics weren't impressed. 'The actors strain for comic effects that aren't there,' Vincent Canby commented acerbically in *The New York Times*. The studio, for its part, seemed more interested in its second *Indiana Jones* movie, which came out a year later.

The Film Commission started to become concerned at the number of films using tax-sheltered financing. In July 1981 Bill Sheat told the government there should be a ceiling on the amount of money that could be used. But in August the following year, when prime minister Robert Muldoon announced new rules for investment in film, he hadn't accepted this suggestion, opting instead to eradicate all tax-shelter structures 'arising from the use of so-called non-recourse loans and other leveraging arrangements'. It was, he said, ridiculous that some people could get tax concessions of up to 150 percent. His associate minister claimed film investors had been claiming even higher tax deductions – up to 180 percent.

The commission was not happy: the new rules were unlikely to attract any private investment. Feature films, Sheat complained, had been singled out for most unfavourable treatment. He feared production would be rapidly reduced. Dave Gibson, president of the Independent Producers and Directors Guild, said 'the industry did not intend to die quietly'.

The Academy of Motion Pictures organised an exhibition at parliament to explain why film-makers opposed the new measures. We all went to the opening to plead the industry's case with suspicious members of parliament. Newspapers warned that the new film industry might be losing its future. 'Generous although past assistance may have been, was the change really necessary?' *The New Zealand Herald* asked.

John Barnett came back from an international film festival 'astonished to find that what people see in a New Zealand film is often the sum total of their knowledge of this country … It seems crazy that the New Zealand government won't provide incentives, as so many other countries are doing.' I agreed. At film markets in the 1980s I was often asked if New Zealand was near Greenland or New Guinea. For many people in many countries, our films were the first opportunity to learn anything about New Zealand.

Alan Highet promised the government would consider other ways of sustaining the film industry if the changes proved as disastrous as film-makers predicted. The short-term compromise was a two-year grace period. The old rules would continue until September 30, 1984. The result was production levels that were never reached again. In the three years from 1982 to 1984, a total of 40 feature films were made in New Zealand – more than half the decade's total production. Titles which came to New Zealand included two more produced by Antony Ginnane, including *Second Time Lucky*, which gave Jon Gadsby the chance to play the Angel Gabriel, with Robert Helpmann as the Devil and Robert Morley as God. Jodie Foster co-produced as well as starred in *Mesmerized*, which was shot in the Bay of Islands.

There's continuing debate as to whether these were good or bad times. I thought it was a great period. The New Zealand industry wasn't swamped by other people's movies. Three-quarters of the 40 productions were created in New Zealand by New Zealanders. There was work and opportunity – and a reasonable ratio of successes, most notably the launching of Vincent Ward's career, which would bring New Zealand into the Cannes competition for the first time. John Barnett analysed the films funded with tax money, and compared them with films financed by the commission. He found that the percentage of successes and failures was the same.

There were disappointments, of course. One was Geoff Steven's second feature *Strata*, which was shot on White Island, an active volcano off the coast of New Zealand, and the volcanic plateau of Tongariro National Park. The screenplay was a psychological thriller co-written by Steven, Michael Havas and a Czech film-maker named Ester Krumbachova.

Krumbachova was listed in *The Oxford Companion to Film* as having made a substantial contribution to the new Czech cinema. She came to New Zealand with Michael when the film was being developed. I drove her to Makara beach outside Wellington, where she wept at the beauty of the yellow gorse-covered landscape. After dinner she sang songs with her guitar and told the story of her life, with Michael translating. But her writing talents did not seem to mesh with a film being made in New Zealand.

During my next overseas trip I added a visit to Prague to see if Michael could help us meet buyers from behind the Iron Curtain. It was an unnerving experience. The atmosphere in Prague was tense and most of Michael's film-maker friends seemed to be out of favour with the secret police. I was shown films by banned directors behind locked and guarded doors. When the time came to leave the country, immigration officials were not willing to let me leave the terminal and walk to the plane. Had they discovered I had been in contact with the wrong people? No, they wanted cash, Michael signalled. I paid up, and ran for my flight.

My unease was well based. I learnt later that the secret police had been watching Havas since he'd travelled to New Zealand with Krumbachova. They had pressured him to inform on Czech émigrés in other countries, which he refused to do. While he was in Frankfurt working on a new film, the Czech foreign police forced him to relinquish his Czech citizenship. He did not regain it until 1991, after he had assembled 17 documents to prove he had been born in Czechoslovakia and thereby satisfied the suspicious bureaucracy.

Not all productions from offshore needed New Zealand tax deals. Expatriate Andrew Brown, whose British television career had included producing *Rock Follies* and *Edward And Mrs Simpson*, returned home with tax-sheltered British money for a feature named *Bad Blood*. The script, which he had written himself, was based on a famous manhunt which had taken place on the South Island's West Coast in 1941 after a farmer named Stanley Graham shot three policemen and a school inspector. Brown produced the film, which was directed in Hokitika by an unknown Englishman called Mike Newell who, 12 years later, would reach international fame with *Four Weddings And A Funeral*.

Australian actor Jack Thompson starred alongside a large New Zealand cast. 'They're not, thank God, that bland kind of cast you often get,' Brown told me, thinking perhaps of his experiences in Britain. 'They're good, raw character actors, all 48 of them.'

Post-production was carried out in primitive conditions in London. When I visited the film-makers in a Soho attic, they talked about the high standard of technical facilities and talent they'd found in New Zealand. 'If only we'd known,' they said. In those days the talents and resources of New Zealanders had not yet earned international recognition.

Bad Blood is now an important title in New Zealand's cinema history but in the 1990s, after Andrew Brown died in mid career, it seemed to disappear from availability. For years we couldn't find out who controlled the international rights, and as a result couldn't include it in New Zealand retrospectives.

The boom period also saw the first feature by a Maori director. Merata Mita's documentary *Patu!* – about the nationwide protests against the 1981 tour of New Zealand by South African rugby team the Springboks – wasn't helped by tax-sheltering investors: its small budget came from religious and anti-apartheid groups' donations, Arts Council grants and a Film Commission loan. Many film industry specialists worked as unpaid volunteers, and the 16 photographers, including Roger Donaldson, were often in danger as they filmed violence between anti-tour protestors, tour supporters and police.

The film was conceived as 25 minutes, but when it was completed after two years in post-production it ran for 110. It had its world premiere at the Wellington Film Festival and was acknowledged as uniquely and controversially important because it gave a Maori critique of the state's heavy-handed response to the protests at a time when documentaries weren't expected to have a point of view at all.

I had never met Merata Mita, and sought her out through mutual friends. One of nine children, Mita had received a traditional Maori upbringing in a small Bay of Plenty town and trained to be a home science teacher. She had started to use film and video with students, and this had led to a job as coordinator on a 1977 documentary on the Treaty of Waitangi. Concerned by what she saw as misrepresentation of Maori views, she had started to make her own documentaries – including some with German expatriate film-maker Gerd Pohlmann – as well as working as a reporter and researcher on a Maori television series.

Over dinner we talked easily about personal subjects unrelated to movies, and after coffee I told her I wanted to help with *Patu's* marketing and promotion. But she was unhappy that the commission had initially turned down her request for financial support and wouldn't accept my offer, instead choosing Jonathan Dennis to represent her film. Jonathan began by writing to Contemporary Films in London, and they became the first of my contacts from the film society years to release a New Zealand film.

The Guardian called *Patu!* 'unquestionably one of the strongest and most vivid documentaries of recent years '.

VIGIL

Ein Film von Vincent Ward

**Vier Menschen
am Rande der Welt.
Und ein Fremder
der Unruhe bringt.**

**FUTURA
FILM**

SEVEN

STEVEN SPIELBERG
SPIES VIGIL

O UR THIRD CANNES FILM FESTIVAL in 1982 was a landmark. Sam
Pillsbury's first feature, *The Scarecrow*, was selected for the Directors
Fortnight – becoming the first New Zealand feature to win official selection
at Cannes.

Sam and his producer Rob Whitehouse had looked outside New Zealand for a
lead actor to play the murderer, disguised as a magician, who arrives in a small town
seeking his next teenage victim. They chose veteran Hollywood character actor John
Carradine, who had appeared in *Bride Of Frankenstein* in 1935 and *The Grapes Of
Wrath* in 1940. Sam met the star at Auckland Airport, and discovered that his hands
were twisted by arthritis. Arrangements for a fight scene had to be reorganised and
rechoreographed. While shooting the film, Carradine celebrated his seventy-fifth
birthday.

By this time Dorothee Pinfold had left the commission to join Dave Gibson at
his fast-growing production company. Her successor as distribution manager was
Judy Russell, one of the founders of Playmarket where, as its first director, she had
pioneered the licensing of New Zealand plays to local theatre companies. One of her
first tasks had been to organise the first visit to New Zealand of Pierre-Henri Deleau,
delegate-general of the Cannes Directors Fortnight. She drove him to the National
Film Unit for a preview screening of *The Scarecrow*. 'I'll take it,' he told her as the
final credits rolled. At first she didn't realise the significance of this.

The Scarecrow was based on a novel by the writer Ronald Hugh Morrieson,
who had died ten years earlier after living his whole life with his mother in a small
provincial town. Deleau said it 'spoke to me very well about a country I don't know.
It had the alchemy which I look for in a movie.' The Directors Fortnight, which
Deleau had founded in 1969 to rival the official competition, showed the film four
times in the Star Cinema, with Sam Pillsbury introducing each packed screening

and answering questions afterwards. Later he would remember the experience as both exhilarating and depressing. Although the audiences were enthusiastic, he worried that his film might not become a box-office hit.

The prestige of Cannes selection in fact helped with sales. I signed two US deals. One was for a theatrical release by Frank Moreno, who had just released the Australian hit *Breaker Morant*, and who decided against our wishes that *Klynham Summer* would be a better name for Sam's film. The second American sale was for telecasting by the Entertainment Channel, a new movie network, whose buyer was Carol Greene. Donna Gigliotti wanted to buy the film too, but her bosses at United Artists Classics wouldn't agree: they had seen powerful critic Rex Reed sleeping through part of the screening.

The Scarecrow's official selection for Cannes gave us our first experience of the demanding profession of subtitling. Jeannine Seawell introduced us to her Paris neighbour Anne Head, an English woman who specialised in the complex task of translating dialogue from its original language into French, and then editing and rewriting the results until they fitted the time and space available on screen. Anne would often work all night, dictionaries by her side, to meet Cannes deadlines. She continued to be the French subtitler of most New Zealand films until she moved to Los Angeles in the 1990s. Then we discovered that Jeannine's daughter Michelle had developed equally impressive skills, and she took over.

I shared another experience of an official Cannes screening with Sam Pillsbury, when Monique Malard organised tickets for us to see Steven Spielberg's *ET*. As we left the cinema while the final credits rolled, two thousand people seemed to be sobbing into their handkerchiefs. We felt we were the only people unmoved.

Soon afterwards *The Scarecrow* became the first New Zealand feature selected by the Edinburgh Film Festival. Later, when the film was released in London, Derek Malcolm wrote in *The Guardian* that it was 'a genuine original … another example of a national cinema with something to say and learning to say it.'

The following year I accepted an invitation to participate in the Manila Film Festival, which had been created by Imelda Marcos. The president's wife wanted the festival to become as important – and as big – as Cannes. I hadn't attended the first one in 1982, when competition screenings (including *Smash Palace*) had taken place in the newly completed Manila Cultural Centre, which had been built in 170 days. Audiences hadn't lingered in the giant-sized auditorium because there was a strong smell of cement dust, and there were rumours that terrorists had planted explosives in the walls.

Smash Palace had won the best actor award for Bruno Lawrence, who had been competing against such international stars as Jeremy Irons in *The French Lieutenant's Woman*. Bruno was late arriving at the awards ceremony. His taxi had crashed in the chaos of Manila's traffic and he had needed medical attention. In spite of his bandages, he celebrated his award by rock 'n' rolling with Madame Marcos, watched by jury members who included Satyajit Ray of India, Krystoff Zanussi of Poland and Gillo Pontecorvo of Italy.

New Zealand was invited to bring a larger contingent to the second festival in 1983. As an enticement the organisers paid for both travel and hotels. As a result 38 countries were represented and a market, where I opened a New Zealand office, was established. The festival accepted our offer of John Barnett's production *Wild Horses* for its competition, and showed *The Scarecrow* and John Reid's *Carry Me Back* in another section.

Every night the First Lady entertained her three thousand international guests, accompanied by the American actor George Hamilton and a retinue of friends known as 'blue ladies'. At one of her dinners, three thousand soldiers acted as waiters so every guest could be served at the same time. At a party in the presidential palace, she sang to us slightly off-key and invited us to admire her rooms full of gowns and shoes. Across the dimly lit palace gardens were high walls, well guarded, and we sensed the crowds of local people outside.

For a festival parade, Madame Marcos equipped Manila's street cleaners and their families with festival T-shirts and paper flags. They were told to line the route and wave as the procession went past. On another occasion she invited guests to join her for a sunset cruise on the presidential yacht. For those who accepted, including most of the Italian delegation, the cruise became nightmarish when she extended it until the next morning because she was enjoying herself so much. Her guests were trapped at sea and there was panic in Manila – and many cancelled business meetings – when no explanation was given for the yacht's failure to return to port.

There was never a budget for the Manila festival: its director was able to draw credit from the Central Bank whenever he needed it. Madame Marcos ensured that the highest levels of society supported her event. Bill Sheat and his wife discovered that their liaison officer was the wife of the air force chief of staff.

No other film event equalled the lavishness of Manila in 1983. Although our movies won no awards, I justified my attendance by the fact that so many film people were there. I returned home with ten more sales of New Zealand films.

Four months later we were back at Cannes, where Geoff Murphy's epic movie *Utu* was selected for the official 'out of competition' section in which *ET* had premiered the year before. A spectacular drama about a Maori warrior seeking revenge after his tribe had been slaughtered by British soldiers, *Utu* screened to more than two thousand people one day before *Merry Christmas, Mr Lawrence*, in the same massive auditorium of the new Palais des Festivals.

'A film at the same time spectacular and intimate … strange and astonishing,' wrote *Le Monde*, while *L'Express* detected 'a primitive lyricism, accentuated by the incredible beauty of the countryside'. 'Is the New Zealand soul a closed secret to us?' *Le Matin* asked, adding: 'Far from being pompous or inflated, the film spares no one, neither the naïve or arrogant Englishman nor the Maori who has read *Macbeth*. Geoff Murphy keeps an ironic distance from the narrative conventions and moral clichés of the Americans, which the Australians do not achieve.'

Actor Anzac Wallace, who played the Maori warrior, told *Le Figaro* the film marked 'the true end of colonialism, the rebirth of a culture'. The interview was one of many organised by Denise Breton, an influential French publicist who had encouraged the *Utu* team to choose the out-of-competition selection instead of the Directors Fortnight which had also invited the film. Her recommendation was energetically opposed by French producer Pierre Cottrell, who was translating the film and creating its French subtitles. But everyone else wanted the prestige of the main programme and the new palais.

We also accepted Breton's recommendation that there should be a lunch for the press at the beach restaurant in front of the Carlton, the famous century-old hotel on the Croisette. We booked the restaurant and sent invitations to the press boxes of the world's best film journalists.

After the morning's media screening, we waited in the sunshine for the journalists to stroll along the Croisette to the *Utu* lunch. As they neared the steps leading down to the beach, they were distracted by the appearance of an American actress on a high balcony of the Carlton across the road. Well known for self-publicity, Edie Williams opened her blouse, revealed her enormous breasts and then disappeared back into the hotel. Less than a minute later she reappeared at the front door, crossed the road and ran towards the restaurant pool, where she removed all her clothes except for a G-string.

Lindsay Shelton collection; photo: John Maynard

Raetihi 1978. *The author visits director Geoff Steven (right) on location during the shooting of* Skin Deep, *the second feature film supported by the interim Film Commission. The film tells the story of the ructions which ensue when a massage parlour opens in a small New Zealand town.*

Cannes 1981. *First Film Commission chair Bill Sheat and his wife Genevieve on their first visit to the world's leading film festival in the south of France. New Zealand features which screened in the market included Roger Donaldson's* Smash Palace, *which launched the director's international career.*

Cannes 1982. *Entertaining influential critics was an important part of marketing New Zealand films. Peter Noble (left) was a high-profile columnist for the British trade paper* Screen International: *Cannes festival-goers vied for mention in his daily column. Philip French (centre), critic for London Sunday paper* The Observer, *had written that New Zealanders wisely restricted themselves to movies 'with modest budgets and indigenous subjects'. The author is on the right.*

Sydney 1982. *Don Blakeney (left), director Geoff Murphy and Kerry Robins deliver the print for the Australian premiere of* Goodbye Pork Pie. *Don Blakeney, the New Zealand Film Commission's first executive director, was now a producer and sales agent. Kerry Robins (right) was another commission staff member who had moved into film production and sales. Murphy would make three more features in New Zealand before re-starting his career in Hollywood.*

Manila Film Festival 1983. *New Zealanders and friends during the extravagant film festival organised by Imelda Marcos. From left: Lena Enquist of the Swedish Film Institute; Keith Aberdein, star of* Wild Horses, *which was screening in the festival's competition; New York distributor Carol Greene, who had bought the rights to telecast* The Scarecrow *in the United States; French sales agent Michel Brodeaux; German sales agent Wolfram Skowronnek; Rob Whitehouse, producer of* The Scarecrow, *which was in the official programme; New York distributor Donna Gigliotti; and the author.*

Cannes 1983. *Anzac Wallace (left) and Kelly Johnson, two actors from Geoff Murphy's* Utu, *sing at a New Zealand reception held to celebrate the film's official selection for the festival.*

Cannes 1984. *Director Vincent Ward arrives for the official festival screening of his first feature* Vigil. *With him is an actress provided by the festival as his companion.* Vigil *was the first New Zealand feature to be selected for the Cannes competition.*

Cannes 1985. *A pause between meetings in the New Zealand office. From left: office manager Monique Malard, the author, Paris sales agent Jeannine Seawell, and New Zealand producer Larry Parr. A prolific producer, Parr had three new films screening in the festival market that year.*

Cannes 1985. *Director Michael Firth (centre) at a New Zealand reception with two Englishmen who had acquired the rights to release his film* Sylvia. *John Hogarth (left) of Enterprise Pictures would release it in British cinemas; the BBC's Alan Howden (right) had acquired British television rights.*

BELOW: Other guests were Canadian Linda Beath (left), a New York-based distributor, Variety *correspondent Mike Nicolaidi and London sales agent Carole Myer. Behind is the Miramar apartment building. In the 1990s the word Miramar would gain fame as the name of the Wellington suburb where Peter Jackson based WingNut Films and built a lavish post-production facility.*

Wellington 1985. *Bill Sheat (left), chair of the New Zealand Film Commission for its first eight years, is farewelled by his successor David Gascoigne, who would head the organisation for the next eight.*

Wellington 1986. *The Film Commission board. From left: producer Larry Parr, retired banker Merv Corner, new chair David Gascoigne, actress Dorothy McKegg, exhibitor Royce Moodabe, producer-director Sam Pillsbury and deputy secretary of internal affairs Brian McLay.*

By now our guests had forgotten lunch. They gathered around the pool to see what would happen next. The blonde actress jumped into the pool and removed her G-string. At this point the crowd was getting out of control. An over-excited young Frenchman jumped into the pool and started to assault the naked American, who kept smiling for the cameras as she was pushed under the water. Pandemonium ensued. The *Utu* press lunch began very late, with very few journalists and much more food for the New Zealanders than we had expected.

Don Blakeney, who had left his job at the Film Commission to produce *Utu*, at first retained the film's international sales rights for himself. He set up an entity called Scrubbs and Co, and helped by Kerry Robins and investor David Carson-Parker began sales at Cannes from a desk we gave him in the commission office. After Cannes, the film was released in France by Gaumont, one of Europe's pioneer film companies. Gaumont had been distributing films since 1896. It advertised *Utu* as 'un western sauvage, epique et flamboyant'. The film screened in 14 Paris cinemas, opening the same week that Roger Donaldson's American production *The Bounty* was launched in 34.

After the festival, Blakeney had re-cut the film, reducing its running time by 20 minutes and reshaping it in a way he believed would be more palatable to international audiences. The director played no part in the re-editing, but his final approval was necessary to satisfy the law of France, where the distributor had to be given a certificate signed by the director authenticating the work.

Utu began its United States release in New York in September 1984 – eight months after *Te Maori,* a ground-breaking exhibition of Maori artefacts, had been launched with fanfare at the Metropolitan Museum. The exhibition had given thousands of New Yorkers their first experience of Maori art and culture, but had not acquainted them with the seriousness of the nineteenth-century conflicts re-enacted so stylishly in the movie. *The New Yorker*'s famously wry film critic Pauline Kael wrote that Geoff Murphy had an 'instinct for popular entertainment and a deracinated kind of hip lyricism'.

Apart from its screenings at Cannes, the longer version of *Utu* was not seen outside New Zealand for 15 years, when I licensed it for DVD release in the United States – complete with composer John Charles' overture, played over a blank screen. Don Krim of Kino International promoted the DVD as 'the unseen original director's cut'.

At the time of the film's New Zealand release, Don Blakeney, expecting record audiences, had been disappointed by the numbers. However I thought it had been a great success. It had attracted more than 250,000 people, and *The Sunday Times* had proclaimed it 'a cultural benchmark'. Don Blakeney never produced another feature film. Four years after *Utu* was launched, he handed the sales rights back to the commission and gave me his files, saying, 'I didn't realise your job was so hard.'

The year *Utu* hit Cannes there were three other New Zealand films in the market. One was Geoff Steven's volcanic thriller *Strata*. Another was John Laing's *The Lost Tribe,* also a thriller, about a man seeking his missing brother in the wilds of Fiordland, a film which went on to win awards in Spain and France. The third, *Wild Horses,* was a western-style drama filmed in Tongariro National Park and directed by Derek Morton. Morton, who had been successful as a television producer and then as a maker of commercials, and had worked on the crew of *Smash Palace*, left the production before it was finished and wanted to remove his credit. He told *Onfilm* magazine he couldn't make head or tail of the story, and he asked me never to mention his name in news releases about the film. But his credit stayed on the film, the producer citing contractual requirements.

A month before Cannes we had participated in a three-week retrospective of New Zealand films at Paris's Cinémathèque Française, a venerable institution which has been called the Louvre and Museum of Modern Art of film. The retrospective was billed as one of the biggest New Zealand events ever presented in Europe: its 24 programmes showed almost every film made in the country in the previous 45 years. Later that year I went back to Paris for a private meeting with the general delegate of the Cannes Film Festival, Gilles Jacob. In his panelled office in the Rue du Faubourg St-Honoré, I told him we had high expectations for the first feature then being completed by 26-year-old Vincent Ward. We believed it could be the first New Zealand film to win selection for the Cannes competition.

Jacob was interested. Early the following year we sent him the first print so he could view it in his private screening room. We also showed the film to Pierre-Henri Deleau when he made his second visit to New Zealand. The film was a masterpiece, said Deleau, and he immediately selected it for the Directors Fortnight. But we were aiming for the top: when Jacob chose the film we turned down Deleau's offer.

At the 1984 Cannes Film Festival *Vigil* became the first New Zealand film to screen in the official competition. In honour of this, the festival offered Vincent three nights' accommodation in an antiquely furnished room at the Carlton Hotel. Vincent didn't want the room – he was comfortable in the backstreet hotel we had found for him – so his producer, John Maynard, and the film's production manager, Bridget

Ikin, decided to occupy it in his place. In the middle of the night John was awoken by a phone call. A sultry female voice asked if he were alone. It was never clear whether this, too, was an offer from the festival.

An hour before the competition screening, Bill Sheat and I gathered in the room with Vincent, John, Bridget and others involved in the making of the film. In accordance with the festival rules, we were all dressed formally – Bridget in a spectacular gown she had borrowed from a friend in London. Vincent had been provided with a young Asian film star to be his companion. When our time arrived we were escorted downstairs, through crowds who gazed at us enviously from behind roped barriers. Limousines drove us the few hundred metres along the Croisette to the new Palais where, watched by scores of paparazzi, we walked up red-carpeted stairs to shake hands with Gilles Jacob.

Vincent's seat of honour in the stalls of the 2,200-seat Louis Lumière Auditorium included a telephone through which he could call the projectionist if he had any concerns during the screening – an unlikely prospect given the French passion for perfection in sound and vision. At the end of the film, spotlights focused on him as the audience applauded.

Although the critics were divided, *Vigil*, with its moody story of a solitary girl growing up on an isolated farm, won plenty of influential admirers. In *The Times* David Robinson called Vincent Ward a rare visionary. *La Repubblica* named him 'the standard-bearer of the New Zealand new wave'. *Le Monde* critic Louis Marcorelles wrote that the film 'leads us into the spirit of silent cinema … and demands total abandon from the spectator'.

Vincent had searched all over the country before choosing a remote and wintry location for *Vigil* at Uruti in rural Taranaki. Some reviewers speculated that the farm sequences were an echo of his childhood. He had grown up on a farm which his father, a third-generation New Zealand farmer, had bought after returning from the Second World War with his new wife, a German-born woman who had left Hamburg at the age of eleven to live in Israel and had joined the British Army at the start of the war.

Jeannine Seawell handled European sales of *Vigil,* and although she could never find a distributor in France she signed deals for the film to be released in England, Italy and Germany. In Germany it screened in more than 50 cities.

When I couldn't find an American distributor, I sought help from Los Angeles publicist Mickey Cottrell, who had worked on Don Blakeney's United States release of *Utu,* and we released the film ourselves. Mickey rented a cinema at the Samuel

Goldwyn Company's new art-house multiplex in a vast Los Angeles shopping mall where *Utu* had played to good houses. *Vigil* received good reviews, with *The Los Angeles Times* calling it 'an extraordinary visual and psychological experience, a work of awesome beauty at once mystical and earthy'. Such praise helped attract bookings through the rest of the United States.

It took me a year to finalise a sale of *Vigil* to Japan after I met Tetsu Fujimura, a young Japanese man who was making plans to set up a new distribution company. At each of three markets I told him about the importance of Vincent's film and its growing international successes. When he started his company and named it Gaga, *Vigil* became one of his first acquisitions.

Steven Spielberg had also heard about *Vigil*. He requested a print for a private screening at Amblin Entertainment in Los Angeles. Did this mean he might ask Vincent to direct a film in Hollywood? We sent a print as requested by the great man but received no feedback, not even a thank-you note.

New Zealand's Cannes representation was growing exponentially. In 1984 it numbered 25, including six producers, five directors and seven people involved with sales. As well as *Vigil*, we screened nine other films in the market, twice as many as in any previous year. Of these films, three had been made solely with tax-sheltered investment, three solely with Film Commission investment, and three with a mix of money from the commission and tax-sheltered investors.

Yvonne Mackay's *The Silent One* had been shot on the exotic location of Aitutaki in the Cook Islands. It was not only Yvonne's first feature film but also New Zealand's first fiction feature directed by a woman. Ian Mune's script about a lonely boy whose friend is a turtle with magical powers was based on a book by leading New Zealand children's author Joy Cowley. The Dolby stereo sound mix was completed only days before Cannes, so producer Dave Gibson carried the first print to France as personal baggage – just as Nigel Hutchinson had done with *Goodbye Pork Pie* four years before.

None of us had yet learned that there were efficient, professional courier systems for delivering prints to markets. When *Strata* was booked to screen at Mifed, John Maynard and Bridget Ikin had carried the print with them on the train journey from London to Milan. En route there was a police emergency when the two metal containers were mistaken for bombs. I had a less serious problem as I struggled to carry two containers of film and my briefcase during a New York rush hour. When I failed to find a taxi, I arrived late for the screening and out of breath. The distributor sympathised with my embarrassment. He gave me a list of phone numbers of courier companies specialising in handling prints. Never again did we worry about personal deliveries.

Dorothee Pinfold, by then handling sales for the Gibson company, was responsible for selling *The Silent One*. She negotiated the highest-yet advance for a United States contract with my friends at Satori. All of us, especially Dorothee, were less impressed when Satori failed to pay all the money. The case ended in a New York court. Gibsons won but the money never arrived: Satori had collapsed when its releases failed to attract the audiences, and the income, it had expected. This proved to be an uncomfortably frequent event in the American film distribution world. We were learning that releasing films, like making films, was a risky business.

New Zealand's 1984 market screenings included four more first features: Bruce Morrison's *Constance*, Lynton Butler's *Pallet On The Floor*, Peter Sharp's *Trespasses* and Melanie Read's *Trial Run*. Read's film was the first production by former sound operator Don Reynolds. We also screened his second production, *Heart Of The Stag*, a Michael Firth-directed drama about incest starring Bruno Lawrence as a loner who rescues a terrified young woman from her father's assaults. *Newsday* likened Lawrence's performance to 'the tough tenderness of a young Brando'.

Also on our schedule was John O'Shea's production *Among The Cinders*. This coming-of-age drama about a 16-year-old boy had been made with finance from Broadbank and the Film Commission, plus a million-dollar investment from North German Television, who had insisted on a German director. Rolf Haedrich had been hired, and had co-written the script with O'Shea. 'The young girls in the film were led along a distinctively European path by the director's misapprehension of the basic story,' O'Shea wrote 15 years later. He felt the Maori content of the original story had been lost. The film premiered at the Berlin Film Festival, and screened on German television as *And He Took Me By The Hand*.

I licensed *Among The Cinders* to New World for the United States. It was their second New Zealand feature after *Battletruck*. To promote the film, they used our New Zealand poster, one of the rare occasions when Americans didn't redesign our advertising.

Altogether I made five round-the-world trips in 1984. The schedule was intense, but the work was exciting as international recognition of New Zealand films kept increasing. In February, after a working week in Sydney, I flew to the Berlin Film Festival for two weeks of New Zealand screenings in the market. I travelled home again through London, New York and Los Angeles, all places where there were a great many distributors who wanted to hear about our films. Each day would begin with a breakfast meeting, followed by two morning meetings, a meeting over lunch, two (sometimes three) afternoon appointments, and of course more business discussions over dinner.

After six weeks at home – which included the New Zealand premiere of *Constance* – it was time for MIP, then private screenings of *Vigil* for French distributors in Paris, followed by two weeks' work at Cannes. After more meetings in London, Paris, New York and Los Angeles, I spent a month in New Zealand and flew back to Europe for screenings of New Zealand films in two Italian festival competitions, with prizes at each. On the way home I stopped in Bangkok to meet the diplomat who had helped promote *Skin Deep* in Paris, and to find out how he could help us in Asia.

Then there were four weeks in Canada for screenings and sales of *Vigil* and *Constance* at the Montreal and Toronto Film Festivals, which were competing for the North American premieres. After five weeks back in the office I returned to Italy for the two-week Mifed market, followed by a week in a freezing November at the Chicago Film Festival, whose founding director Michael Kutza was another supporter of New Zealand cinema.

It had become clear that festivals were the best way of introducing New Zealand films to the world – especially festivals with competitions, where we could attract extra attention by winning awards. As well as the big three – Cannes, Berlin and Venice – there were many other festivals, some huge, some small, with authoritative directors, juries who seemed to like our films, and supportive media who were interested in writing about new film-makers. Some of the festival representatives came through the door of our market offices to ask about our films, others had to be persuaded to take an interest, and some – such as Venice – took years before they showed their first New Zealand movie.

New Zealand diplomatic posts were always happy when our films were shown in the countries where they were based. When five of our films were in competition at the Knokke-Heist Film Festival in Belgium, the New Zealand embassy in Brussels organised a reception. Film festivals in Hong Kong, Jakarta, San Francisco, Karlovy Vary, Vienna and other places introduced their audiences to the hitherto-unknown islands of New Zealand.

The Scarecrow became the first New Zealand film to win an award in Italy when it scored best ensemble acting at Mystfest in Cattolica, one of many summer film festivals supported by small European cities. *Beyond Reasonable Doubt* won New Zealand's first French award at the Cognac Festival of Thrillers. *Pictures* was the first New Zealand feature selected by the Berlin Film Festival, and the first to win an award at the Moscow Film Festival. A year later *The Silent One* also won at Moscow, and at film festivals in Frankfurt, Paris, Giffoni in Italy, Figuera de Foz in Portugal, and Chicago. *Vigil* won best film at Madrid, and was chosen as the most popular film at a festival in the small French town of Prades.

While festivals proved a useful mechanism for getting our films in front of foreign audiences, we soon discovered the enormous power of television. Our sales to Klaus Hellwig resulted in a top-rating season of New Zealand movies on ZDF, one of the biggest German television networks. Hellwig had also directed a documentary about New Zealand, and this was telecast in the same week. 'New Zealand is larger than West Germany but has a population of only three million people,' the narrator explained.

The ZDF season generated considerable interest. German critic Hans Blumenburg wrote in *Die Zeit*: 'In the culture pages of European newspapers there has been no place for New Zealand until now.' The film industry, he said, was the only recent development in New Zealand that had been noticed by the world. He was right. Inquiries at the New Zealand government's tourist office in Frankfurt doubled, reinforcing our claim that the film industry added value to other sectors of the economy. The office hadn't been carrying out any advertising at the time.

The BBC, too, ran a first-ever season of New Zealand feature films. This opened with *Pictures* and included *Middle Age Spread*, *Solo*, *Skin Deep* and *Among The Cinders*. Each film was seen by more than three million British viewers. We would license more than 20 features to the BBC over the next 20 years, all through Alan Howden, encouraged and advised by programmer Brian Baxter. When Baxter came to New Zealand on a lecture tour for the British Council in 1984 he was able to seek further films for the BBC, as well as propounding his views to university students. In his first lecture to film students at Wellington's Victoria University he extolled the virtues of indigenous cinema and advised: 'Don't sell out like some Australians.'

Baxter was a fervent admirer of Bruce Morrison's *Constance*, which the BBC had bought on his recommendation. I arranged for him to have dinner with actor Jonathan Hardy, who had co-written the script and could vividly re-enact all the roles.

Jonathan had recently had heart bypass surgery. Baxter's partner in London was soon to have the same operation. The dinner became an extraordinary event as Hardy demonstrated the power of primal screams in aiding recovery from major surgery. *Constance* went on to earn high ratings when it was telecast on BBC2.

Britain's new Channel 4 followed the BBC's lead and announced its own season of New Zealand films. Derek Hill, who had closed his New Cinema Club, had the job of buying films for the new commercial public channel. I sold him *The Scarecrow*, *Sleeping Dogs* and *The Lost Tribe* and he bought *Beyond Reasonable Doubt* from its London sales agents.

Back home, Television New Zealand had run its first season of New Zealand films in 1981. With only two television channels in existence, *Beyond Reasonable Doubt* had attracted 38 percent of the population and *Middle Age Spread* 30 percent. The season also included John O'Shea's 1966 musical feature *Don't Let It Get You*, which had been ignored by local television for 15 years and never telecast before. All seven features attracted at least double the ratings achieved by the programmes they replaced, reported the somewhat surprised broadcaster.

These ratings put me in a strong position to begin negotiations for a second and then a third season of feature films. The negotiations always seemed to happen overseas. 'Anywhere, anytime' was my arrangement with head of programming Des Monaghan. We would meet for lunch once a year at one of the international markets and talk about licence fees and titles. It seemed easier to meet overseas than at home. This was also the experience of film-makers, who felt shut out by the broadcaster's policy of producing most local programmes in-house. 'You couldn't get meetings in New Zealand so you'd go to Cannes and see them in the corridors and try to do deals with them,' Dave Gibson recalled.

For the top New Zealand features, Television New Zealand's licence fees would, by the end of the '80s, peak at around $100,000 – five times the rate we had first been forced to accept. Given the films' high ratings, however, this figure seemed a modest contribution to the cost of production. In the '90s the state-owned broadcaster seemed to forget local films could be so popular, and licence fees slumped.

After the departure of Don Blakeney to pursue his new career as a producer, the Film Commission was without an executive director for two years. Jock Maclean was hired as financial and projects director, and until 1983 he and I shared administrative duties. Then Jim Booth, who had handled the affairs of the interim commission while at the Internal Affairs Department, left his job as assistant director of the Arts Council and began a new life as the commission's second executive director.

Jim brought new initiatives, including a popular scheme for financing short films, which he branded 'bonsai epics'. He and I had an edgy relationship, each accusing the other from time to time of lack of consultation. My overseas trips meant I was often absent when policy decisions were being debated. My relationship with Don had always involved consultation before decisions were reached, but Jim's attitude often seemed often to be 'out of sight, out of mind', as he once told a colleague when I was in Europe.

Some of Jim's other relationships were similarly volatile. He once threatened to walk out of a board meeting at which Dave Gibson had criticised him. Gibson told him that as an employee of the commission he didn't have any right to leave. Then Gibson walked out instead.

Jim and I battled over his decision to pull down walls and rearrange the commission's new Courtenay Place office in an open plan. My office schedules were filled with meetings and phone calls, usually involving prolonged negotiations with producers and distributors. I didn't want to do this work in a large space shared with other people.

It took a year before I could get my walls back. By then Judy Russell had decided not to restore the huge wall chart from the commission's first office. On it we had listed each feature and short film we were selling, and tracked each sale and each festival invitation. It had been an impressive display, especially useful for showing visitors seeking to learn about the new film industry. But its time was over: we were now handling more titles than could fit on the wall.

Judy's calm exterior was ruffled only once, when we chose a basement nightclub for the 1984 Cannes reception in honour of *Vigil* – the only time we moved from the Plage L'Ondine. We decided a new venue was needed to mark New Zealand's first participation in the competition, and the club's plush scarlet wallpaper appealed. In the event, the club and the party were more popular than we anticipated. In spite of two bouncers, it was gate-crashed by so many uninvited guests that Judy swore never to return to Cannes. But she relented, and stayed with the commission for another three years.

The year after the gate-crashed party, New Zealand films would earn more substantial sales and distribution and acclaim than they had ever achieved before.

"THE BEST SCIENCE FICTION FILM OF THE '80s."
—Kirk Honeycutt, L.A. Daily News

"An eight-plus. Definitely worth seeing, mulling over, and then seeing again."
—Gary Franklin, KCBS-TV

"A classy, compelling, and powerfully moving sci-fi epic."
John Corcoran, KABC-TV

"As thought provoking as any film this year."
Michael Dare, L.A. Weekly

THE QUIET EARTH

PAVILION CINEMAS

EXCLUSIVE ENGAGEMENT
DAILY: 1:15, 3:45, 5:45, 8:00, 9:55

Westside Pavilion 213/475-0202 10800 Pico Blvd. • Unlimited Free Parking

EIGHT

BRUNO LAWRENCE AND THE END OF THE WORLD

*C*ANNES IN 1985 SET A RECORD. New Zealand screened ten new features in the market, a number not exceeded since. It reflected the last phase of the tax-sheltered investments that had ended nine months earlier.

None of the New Zealand features was in an official section, but it was a great year: Geoff Murphy's extraordinary science-fiction epic *The Quiet Earth*, starring Bruno Lawrence as the only man left alive in the world; Michael Firth's *Sylvia*, the story of radical New Zealand schoolteacher Sylvia Ashton-Warner – a much better portrayal than the 1961 Hollywood version *Two Loves* (released in Britain as *The Spinster)* starring Shirley MacLaine; and Gaylene Preston's ebulliently scary first feature *Mr Wrong*, about a woman who gets much more than she bargains for when she buys a second-hand Jaguar car.

Our New Zealand office was in an apartment on the first floor of the Résidence Festival at 52 La Croisette. This building became New Zealand's regular Cannes address, although we were never again able to rent the first-floor apartment – even when, as was usually the case, it remained empty during the festival. Its owner lived in the Middle East. When Monique tracked him down on the phone, he said our rental payment wouldn't cover the cost of his cigars.

The apartment was magnificently positioned dead-centre on the beachfront boulevard, where thousands of industry people walked all day and most of the night. There was a big grassed terrace at the front where film-makers and guests could sit and talk, and from where we could look down on the street and inveigle buyers to come inside. And there were plenty of buyers. We lured them with a poster announcing '36 New Zealand films', the total number of features now listed in our catalogue.

Three of the 1985 films had been produced by Larry Parr, whose thriving production company Mirage Films had expanded to establish its own sales company. Parr had hired computer whiz Paul Davis from the entrepreneurial world of university arts events, and industry veteran Bill Gavin (who had sold films for legendary producer and impresario Lew Grade and Goldcrest Films in London) as consultant.

Two of Parr's three films would become hits, one in terms of local popularity, the other for its international earnings.

With *Came A Hot Friday*, director Ian Mune delivered his most popular film, a period comedy which would attract more than 300,000 New Zealanders in its local release later that year. Its two small-town con men were memorably played by Peter Bland and Phillip Gordon, with an equally memorable performance by Maori comedian Billy T. James as the Tainuia Kid. This was the third and most successful of the features adapted from novels by Ronald Hugh Morrieson.

Shaker Run, directed by Bruce Morrison, with American actor Cliff Robertson and London-domiciled New Zealander Lisa Harrow, featured spectacular stunts and car chases; its Pontiac Trans-American sped off from Dunedin and raced through the country in the opposite direction to *Goodbye Pork Pie*'s Mini. Substantial income from new video markets made *Shaker Run* New Zealand's top earner at Cannes, closely followed by *The Quiet Earth*.

Dorothee Pinfold, in her second year at the Gibson Group, was selling *The Quiet Earth* for producers Sam Pillsbury and Don Reynolds. She negotiated a United States release with a new company called Skouras Pictures, who impressed us by pushing the film's box office to over US$3 million, qualifying it for *Variety*'s top ten foreign releases of the year, along with *Room With A View*, *Mona Lisa* and *My Beautiful Laundrette*. Perhaps we shouldn't have been surprised. *The Los Angeles Daily News* said the movie was the best science-fiction film of the 1980s and Sheila Benson in *The Los Angeles Times* called Bruno Lawrence 'an electrifying screen presence'.

Dorothee and I had worked late one night to devise a campaign for *The Quiet Earth*. We looked at every one of the film's hundreds of still photographs before choosing an image of Bruno with a gun. The American distributor's advertising agents had a better idea. They ditched our campaign, replacing it with an image which we hadn't considered because it didn't exist as a still. They chose a frame from the last image in the film, showing Bruno alone on a beach with a sky full of planets. It was an inspired choice. The American campaign was used in every country where the film was released.

Skouras Pictures' owner, Tom Skouras, told me about growing up in Hollywood, where his uncle Spyros had run 20th Century Fox and been responsible for introducing CinemaScope and greenlighting *The Robe*, the film which had been a memorable part of my teenage film-going. While I had been watching 20th Century Fox movies on New Zealand screens, Tom had been on the back lot watching the films being made. He became a generous host, introducing me to Venice Beach's Market Street restaurant, where co-owner Dudley Moore occasionally turned up to play the piano.

Skouras Pictures also bought the rights to New Zealand's first horror film *Death Warmed Up,* another of our 1985 Cannes titles which was in demand from international distributors. After making *Angel Mine* six years earlier, director David Blyth had spent time in Europe. In Paris he'd met Alejandro Jodorowsky, the Mexican director of *El Topo,* who had read his tarot cards and told him to go home and make films. The result, *Death Warmed Up,* was the first New Zealand film selected for the Paris Festival of Science Fiction and Fantasy Films, where it won the Grand Prix from a jury that included Jodorowsky.

The film had not been popular in New Zealand, where audiences were often hypercritical of local movies. It hadn't been helped by being released before its international successes had begun, and, as a pioneer of the genre, being burdened by local expectations that horror would always have an American accent.

After I signed the deal with Skouras Pictures for *Death Warmed Up,* producer Murray Newey sent the company the only negative, from which it made prints. A decade later, when Skouras Pictures' contract for *Death Warmed Up* had expired and the company had been liquidated, we faced an unwelcome problem when we couldn't trace the negative. This ended our ability to make new prints, a problem compounded by the fact that all the surviving prints had been cut by various censors. We kept up the search year after year, but never solved the unhappy mystery of the negative's disappearance.

In 2001, my last year at the Film Commission, I received a phone call from Los Angeles. Tom Skouras was forming a new distribution company and was interested in finding out about the latest New Zealand films. I told the caller we would be very willing to do more business with Skouras, but only if our missing materials were found. No one called me back.

Four years later, while writing this book, I found myself wondering again what had happened to the negative. I picked up the phone and rang Tom Skouras at his new company. He agreed to search through old files, and a few weeks later sent details of five Los Angeles facilities with which his former company had dealt in the '80s and '90s. I sat down and started to make phone calls. On my third call I discovered that the original video masters were residing in the vault of a video company. But the location of the film's negative remained a mystery.

Mr Wrong and *Sylvia* were my personal best sellers at Cannes in 1985. Quartet Films, an American distributor whose stylish campaigns I had admired for years, bought *Mr Wrong*. Neither Gaylene Preston nor her producer Robin Laing were happy when the Americans retitled it *Dark Of The Night*, but *The Village Voice* cheered us up with its review: 'Witty and understated. Unlike so many movies in which women get victimised by the script and the director, this has them joining forces and fighting back on their own terms.'

Sylvia, with British actress Eleanor David in the title role, was bought for MGM-UA Classics by Carol Greene. 'Enchanting, edifying and imaginatively erotic,' wrote Carol's former professor Andrew Sarris, reviewing the film in the *Voice*. Sarris always seemed to praise the films his ex-student had bought. For *Sylvia*'s North American release, which began in the New Directors New Films festival, the roaring MGM lion was spliced on before the New Zealand credits. MGM's trailer also began with the iconic lion, and Carol let us use the same opening sequence for distributors all over the world, perhaps giving some the impression this was an American studio production.

Our 1985 Cannes line-up also included Levin film-maker Mike Walker's modestly moving film about Polynesian teenagers, *Kingpin*, starring the charismatic Mitch Manuel. We screened *Should I Be Good* and *The Lie Of The Land*, two meandering features produced and directed by Grahame McLean, one of the country's most experienced production managers, who briefly transmuted himself into a producer-director-writer with the help of investors with tax deductions to claim.

Another of our premieres that year was *Leave All Fair*, the story of expatriate New Zealand writer Katherine Mansfield. The film had been developed and directed by Paris-based New Zealander Stanley Harper, who had also been named as director. A fortnight before the shoot, producer John O'Shea had lost confidence in Harper and replaced him with John Reid, who was flown from Wellington to France to rewrite the script and direct 81-year-old Sir John Gielgud as Mansfield's controversial husband John Middleton Murry. 'A strange unit of New Zealanders, struggling manfully with a French crew,' Gielgud would confide in a letter to actress Claire Bloom. Although the film is an understated elegy to New Zealand, with Jane Birkin as Mansfield saying 'I have a perfect passion for the island where I was born,' it was shot entirely in Normandy.

At the end of the year there were five New Zealand features in the London Film Festival, two produced by John O'Shea. Sir John Gielgud was in the audience for

Leave All Fair, and Barry Barclay's documentary feature about plant genetics *The Neglected Miracle* was introduced by the botanist David Bellamy as 'an epic documentary'.

We were proud that eight New Zealand features from Cannes had been sold for theatrical release in the United Kingdom. When *Came A Hot Friday* opened in London at the Curzon Cinema in Mayfair, London entertainment weekly *Time Out* declared New Zealand had finished the year on a cultural note: 'After Keri Hulme's Booker Prize and Kiri Te Kanawa's chart-busting album comes *Came A Hot Friday*, a splendidly engaging kiwi comic western.'

When *Constance* opened on three London screens, *Time Out* was again enthusiastic, praising the 'magnificently realised performance by Donogh Rees'. Another British reviewer called her New Zealand's answer to Meryl Streep.

Annie Whittle, who had originally been cast as *Constance,* was praised for her role as a woman under threat when *Trial Run* was released in London. Philip Strick wrote in *Films and Filming*: 'Blonde actress Annie Whittle, unknown in Britain but seemingly all over the place in New Zealand as a singer, composer and general phenomenon, looks both authoritative and scared stiff with equal appeal and would be no hardship to encounter more frequently.'

At home, though, the news was less salubrious. With tax deals well and truly ended, production levels had plummeted and momentum had been lost. The commission's 1986 annual report stated that it was now all but impossible to get private investment for New Zealand films. There was also a secondary reason for the slump: the film community was exhausted after such a rapid rate of production.

The fall-off wasn't, however, visible outside New Zealand. We seemed to be selling films everywhere. When I licensed *Sleeping Dogs* to Argentina, it was my fortieth sale of this title, which had become a New Zealand classic in less than five years. Dorothee was first to achieve a sale to China, with *The Silent One*.

By the end of 1986 more than 20 New Zealand features had been shown in cinemas in the United States. Only five years earlier there had been none. After *Sleeping Dogs* and *Smash Palace*, Americans had been able to see (and read about) *Among The Cinders, Beyond Reasonable Doubt, Battletruck, Bridge To Nowhere, Carry Me Back, Constance, Death Warmed Up, Goodbye Pork Pie, The Lost Tribe, Heart Of The Stag, The Quiet Earth, Race For The Yankee Zephyr, Savage Islands, The Scarecrow, Shaker Run, The Silent One, Skin Deep, Squeeze, Sylvia, Utu, Wild Horses* and *Vigil*.

Donna Gigliotti had bought *Came A Hot Friday* for Orion Classics, and Linda Beath's Spectrafilm had bought *Queen City Rocker*. There were plenty more to come.

It was also a record year for theatrical releases in France. *The Quiet Earth* opened in twelve cinemas in Paris, three on the Champs Elysées. I walked along the famous boulevard photographing posters advertising the film as *Le Dernier Survivant*. *Leave All Fair* (French title *Souvenirs Secrets*) and *Constance* each played in five Paris cinemas. Also released in France were *Death Warmed Up*, *The Silent One* and *Other Halves*.

Gaylene Preston flew to Paris for a screening of *Mr Wrong*, the first New Zealand film selected for the Creteille Festival of Films by Women. In midsummer there was a day of New Zealand movies on the giant screen of Paris's famous old movie house, l'Escurial Panorama. It began with *Sleeping Dogs*, followed by *Death Warmed Up*, *Goodbye Pork Pie* and *Utu*. The next year, New Zealand signed a co-production treaty with France, although it would seldom be used because there was little synergy between French and New Zealand film-makers.

New Zealand films kept winning international recognition at a dizzying speed. *Vigil* was the most popular film at the Sydney Film Festival. *Among The Cinders* won an acting award at Karlovy Vary in Czechoslovakia. *The Quiet Earth* won best director awards in Madrid and Rome. *Kingpin* won a special prize at Moscow; the citation praised its story of Maori teenagers for 'the clearest embodiment of the ideas of anti-imperialist solidarity, peace and friendship'. *Sylvia* was named by Andrew Sarris as one of the year's ten best films, in a list that included John Huston's *Prizzi's Honour*.

In 1985 I made plans for a week-long New Zealand Film Festival in Tokyo's Science Hall, with a reception at the New Zealand Embassy. John Reid, now president of the Guild of Film and Television Arts, introduced *Leave All Fair*. The event was reported in 60 Japanese publications, with *The Japan Times* announcing that New Zealand had a staggering degree of creative talent.

Mirage's head of sales Paul Davis and I visited all the Japanese theatrical distributors, the same ones with whom I had had no success five years earlier. This time the festival and the reception and the media enthusiasm encouraged them to start buying our films. We signed contracts for four titles to be released in Japan.

The biggest 1985 retrospective was held at the Kennedy Centre in Washington and entitled 'New Zealand Cinema and its Directors'. The 30-day season was created by Ken Wlaschin, who had moved from the British to the American Film Institute.

New Zealand ambassador Bill (although he preferred to be called Sir Wallace) Rowling gave a reception, and John Reid and Michael Firth came to Washington to introduce their films. Wlaschin and his colleague Suzanne McCormick wrote in the catalogue that New Zealand films had achieved a high standard with very little money and no movie tradition. New Zealand cinema was 'one of the wonders of the world … a major achievement that deserves wide recognition'. We were still using that quote 15 years later.

After smaller New Zealand seasons in San Francisco and Mill Valley (near the home of Francis Ford Coppola), we decided on a further boost to New Zealand's visibility by organising a 'Focus on New Zealand Cinema' at the first film festival being held in the wealthy California city of Santa Barbara. The festival invited lots of New Zealand guests. Geoff Murphy and Merata Mita were there with their baby. Bruno Lawrence charmed everyone, as he always did. Peter Wells arrived for screenings of his 50-minute drama *A Death In The Family,* which he had co-directed with Stewart Main. It was praised by Americans as one of the first fiction films about the effect of Aids.

My arm was in plaster. I had broken it in a fall on ice at the Berlin Film Festival. Berlin had been memorable for another reason: I had met the great French actress Jeanne Moreau, whom I had admired since I saw her in *Nathalie Granger.* She was starring in a feature produced by Klaus Hellwig, who introduced us on a staircase at a festival reception. She told me she would love to visit New Zealand. Star-struck, I promised to look after her when she came. Lamentably, the visit never happened.

A reduced version of the Washington season was presented in Australian cities by the Australian Film Institute. In *The Australian Times On Sunday*, Anna Maria Dell'oso wrote, 'Our neighbours across the Tasman seem to have produced more quality cinema in a shorter time and with a smaller population and market than the Australian industry.' Bill Gosden wrote that the influx of New Zealand stories on to local screens had constituted a major shift in popular consciousness. In the past, the notion of New Zealand stories on the big screen had seemed immodest, or even embarrassing.

Just how much we'd got over this was shown by the ratings earned by TVNZ's second season of New Zealand movies. More than 1.5 million people watched *Goodbye Pork Pie* when TVNZ telecast it in 1985. This 50 percent rating was the highest for any movie since television began in New Zealand 25 years earlier. Second in popularity was *Smash Palace*, viewed by 870,000; then came *The Scarecrow*, with an audience of 740,000.

In the same year as these top-rating successes, the Labour government of David Lange acknowledged the slump in film production that had resulted from axing the tax breaks. It also accepted arguments that feature films could promote trade and tourism, and gave the commission $4.5 million for the 1985–86 financial year – $3.4 million more than ever before – plus $750,000 to establish a fund for short films, which would be a new gateway for young film-makers to prove what they could do. But the industry was slow to emerge from the doldrums: only three features went into production in 1985, and two in 1986.

Things weren't helped when the Inland Revenue Department began an investigation of tax returns relating to 60 film projects, including *Utu*. All the films had been financed before the rules were changed in 1984. It had been a period, producer John Barnett said, when people invested in good faith, under the rules applying at the time.

With producer Pat Cox, Barnett formed Magpie Productions to make a feature film which would be one of the few not to face problems raising money. *Footrot Flats*, based on some of the country's most recognisable cartoon characters, was New Zealand's first animated feature. It attracted equity investment of $5 million from 600 investors, including the publishing company which ran the cartoon strip in its newspapers. Murray Ball, who had created the cartoon strip, directed the film and wrote it with fellow cartoonist Tom Scott.

Released during the 1986 Christmas holidays, *Footrot Flats* was seen by a New Zealand audience of more than 730,000 and grossed $2.7 million, beating the record set by *Goodbye Pork Pie* five years before.

Two songs written and performed by Dave Dobbyn – 'Slice Of Heaven' and 'You Oughta Be In Love' – rose to the top of the music charts. At a branch of the Bank of New Zealand, twelve staff dressed as black-singleted farmer Wal to mark the film's opening day. *The Otago Daily Times* organised a look-alike competition featuring Cheeky Hobson, Wal's girlfriend, and reported that the winner was 'able to balance a glass on the appropriate bits'.

Footrot Flats was even more successful in Australia, where it grossed over A$4 million. *Variety* critic Mike Nicolaidi admired its blend of 'whimsy, warm irony and observation of human and animal frailty'.

The film was launched in the Cannes market the following year. Although it won first prize at the Los Angeles International Animation Festival, its American-based sales agents never found a US distributor. They told me that animation was out of fashion at the time.

There were four new features in the Cannes market in 1986. Two were directed by John Laing: his third feature *Other Halves* ('A remarkable film,' said Brian Baxter of the BBC) and his fourth, *Dangerous Orphans.* To boost the numbers, I organised repeat screenings of titles from the previous year, but I discovered Cannes was interested only in new titles. No one wanted to know about 'old' films.

We also screened Ian Mune's teenage action adventure *Bridge To Nowhere,* which achieved substantial video sales, and Richard Riddiford's romantic comedy *Arriving Tuesday,* which seemed 'too small' for international release. I felt *Arriving Tuesday* was under-rated by everyone. Locals didn't attend, and international buyers didn't buy it. Its problem was that it was a romantic comedy, and no one wanted romantic comedies unless they were from Hollywood with Hollywood stars.

The star of the film, Judy McIntosh, arrived in Cannes late one night after our office was closed. Exhausted after a long flight from New Zealand, she went to sleep on the floor outside the office door. The occupant of a neighbouring apartment woke her up and provided a room for the night.

The growth of video markets helped us earn high prices from new distribution companies, including some cheerful Scandinavians who paid a record licence fee for *Dangerous Orphans*, which Dorothee Pinfold sold them for all of Europe. After Cannes, the film sold out at two screenings at the Munich Film Festival. 'Not even the soccer game between Germany and France, telecast at exactly the same time, could keep the crowds away from the new film from their favourite New Zealand director,' festival director Eberhard Hauff told me.

Charles and Kitty Cooper of Contemporary Films decided to acquire British rights to *Other Halves* because they were interested in its depiction of class and racial conflict. *The Times*' critic, David Robinson, confessed to being sceptical about the story but fascinated to see 'the screen debut of a city, Auckland, which appears as a messy, vital, multi-layered metropolis'.

The BBC presented its second season of New Zealand films in 1987. An audience of over 5.1 million people watched *Came A Hot Friday* and 4.5 million *Constance,* which *The Radio Times* described as 'supremely elegant'. In the same month, *Beyond Reasonable Doubt* attracted 5.5 million viewers for Britain's Channel 4 – its highest rating of the week.

In 1987 I made my second visit to the Sicilian town of Taormina, where Guglielmo Biraghi was directing a film festival which had become influential throughout Europe. He had first invited me three years earlier, when he selected *Constance* for the Taormina competition. In the cliff-top seaside town I had spent eight hot summer days in the relaxed company of film industry friends including Jeannine Seawell, who attended the festival every year, whether or not she had a film in the competition.

The awards ceremony was held in a 2,000-year-old Roman amphitheatre with Mount Etna as a fiery backdrop. *Constance* won the bronze charybdis, which I dropped when running across the marble floors of Rome airport to catch my plane back to New Zealand. It didn't break, but the resulting crack was taken to be proof that the charybdis from Taormina was an antique treasure, as well as an important award.

Ngati, the first dramatic feature written and directed by Maori, had premiered in Critics Week at Cannes, where it had been represented by director Barry Barclay, actor Wi Kuki Kaa and scriptwriter Tama Poata. When Biraghi selected it for the 1987 Taormina competition, I knew the festival would be a valuable showcase. But I also knew that the daily routines at the festival were extraordinarily relaxed – mornings on the beach, lunch on the terrace, afternoons by the pool – and hard to justify as work. So I told the commission I would pay for my own ticket and count the days as holidays, although I would also look after the film and its promotion.

In the informal atmosphere I dined several times with the jury chosen by Biraghi. It included Russian director Nikita Mikhalkov, German director Werner Schroeter, African director Souleymane Cisse and British writer Angela Carter. When they chose their awards they gave *Ngati* first prize, praising its story of community survival in a remote Maori settlement as 'a film of intense poetry with an original and strikingly effective film language'. The gold charybdis was presented to Wi Kuki Kaa, who had given a powerful performance. A critic in *La Sicilia* wrote admiringly, 'If he was not given away by his slightly almond eyes, he could be mistaken for a sun-tanned Sicilian.'

Again I carried a charybdis home to New Zealand. Producer John O'Shea told me: 'You are the only person I know who would buy his own ticket to win a prize for my film.'

Ngati had its North American premiere at the Toronto Film Festival. Back home, it was the only New Zealand film of its year selected for the Auckland and Wellington Film Festivals. Then its theatrical release began at Waipori Bay near Ruatoria, where it had been filmed. Altogether it would be seen by more than 40,000 people, and be chosen as one of the best films of the year by *Metro*, the *Listener* and *The New Zealand Herald*.

At the Film Commission, deputy chair David Gascoigne, now a senior law partner, became the second chair. He continued the close relationship with the staff, and like his predecessor Bill Sheat was frequently in the office. Soon after his appointment, the commission began worrying about low attendances at some New Zealand-made films. Too many local releases had seemed under-supervised, under-financed and lacking in drive and commitment, with some producers of tax-deal films lacking confidence in their productions. John O'Shea was one who sometimes seemed insecure about organising local releases. *Pictures* didn't screen in New Zealand until two years after its British release. *Among The Cinders* wasn't submitted for censorship approval for four years. Larry Parr's production *Pallet On The Floor* wasn't released until three years after its disappointing premiere in the Cannes market. Grahame McLean's two features were barely released at all.

Some releases were hindered by the cinema chains. The producer of *The Lost Tribe* complained about Amalgamated Theatres' abrupt cancellation of bookings. Confirmed bookings for the Wellington release of *Ngati* were cancelled at the last minute by Kerridge Odeon, who then refused this award-winning film any release in Auckland's important Queen Street cinemas. Some kind of industry prejudice against local product seemed obvious.

New Zealand's first film awards, named Goftas after the industry-run Guild of Film and Television Arts which established them, had been launched at a televised ceremony in 1986, merging with television awards already in existence. After its popular success with New Zealand audiences, *Came A Hot Friday* won best film for Larry Parr, best director for Ian Mune, best actor for Peter Bland, and four other prizes including best supporting actor for Billy T. James. *Vigil* won three awards.

The ceremony was followed by a seminar on international film financing and co-production. New Zealand's co-production agreement with Australia was signed later in the year.

At the second Goftas in 1987, *The Quiet Earth* won eight awards, including best film, best director, best actor (Bruno Lawrence) and best editor (Michael Horton). The 90-minute televised ceremony was, however, a shambles, with some winners being told to sit down and not collect their awards because time was running out. American presenter Leeza Gibbons from the American television show *Entertainment This Week* and British actor John Inman from the British television comedy series *Are You Being Served?* were heckled when they tried to announce two awards; they later blamed 'alcohol and boredom'. TVNZ apologised, held a closed investigation, and blamed inadequate planning and communication failures. The awards would continue to be contentious up to the present day.

Overseas, a decision to extend New Zealand's marketing activities at Cannes led to a meeting with Prince Charles and Princess Diana. Paul Davis and I had decided that Mirage and the Film Commission would share the cost of a New Zealand stand in the basement market area of the Palais des Festivals. It was the first time either of us had booked space in this unpopular place. Why would anyone want to work underground, when we could have a view of the Mediterranean from our office windows? But we wanted to check if a basement stand could bring us new business.

Halfway through the festival, the market organisers asked us to come to our stand for a royal visit. The building had been closed for the occasion, and in spite of our official badges the security men wouldn't let us in. Eventually we managed to find an unguarded back exit, and ran through back corridors just in time to see the prince and princess strolling down one of the long aisles, looking at posters and brochures. They then turned right into an area notorious for its grouping of porno production companies. They emerged quickly, blushing slightly, and continued their progress towards us.

Perhaps relieved to see the familiar name of New Zealand, Prince Charles paused at a poster for *Ngati* and asked if Wi Kuki Kaa was a singer. I wondered if he was thinking of Inia Te Wiata. Princess Diana's eyes were downcast throughout. She seemed extraordinarily shy.

John Barnett had refused to allow his productions to be promoted on the New Zealand stand, but offered the chance of a royal handshake he accepted the invitation to join our line-up. As the couple came closer, John produced a poster for *Footrot Flats* from under his coat and pinned it up just in time for the royal arrival. We took it down after they departed.

During 1987, when Vincent Ward and his producer John Maynard were unable to raise private investment for his second feature *The Navigator,* Vincent left New Zealand and moved to Sydney. With *The Navigator* the commission had made its biggest-ever financial commitment, and the critical success of *Vigil* had encouraged a number of international distributors to sign pre-sale agreements to help finance the film. However the film also needed private investors, and when Maynard cancelled the shoot a month before it was due to start he blamed the new tax laws. The film industry had, he charged, been singled out for discriminatory treatment, while activities such as goat-farming and bloodstock continued to use tax shelters. Media reports said private investors had put their considerable stash into a Broadway musical instead.

Gilles Jacob, director of the Cannes Film Festival, sent me a letter in support of the film. 'You know that I would not in any way interfere with your domestic problems,' he wrote. 'But I feel free to ask you, as a friend of cinema, that everything should be done to help Vincent make his second film. He is one of the most promising and original directors of his generation.'

Within a year *The Navigator* had been re-structured as New Zealand's first co-production with Australia, and filming had begun in New Zealand. Jacob wrote to me a second time, asking us to let him preview the film for Cannes.

BAD TASTE

"Will do for video what 'ROCKY HORROR' did for midnight shows."
—Tony Timpone, FANGORIA

GUARANTEED TOP RENTER
MAGNUM Entertainment

"This year's Re-Animator"
—GOREZONE MAGAZINE

"I've never seen a movie that's so disgusting— It's great!"
—HOLLYWOOD REPORTER

SUGG. RETAIL: $79.98 • COLOR
PREBOOK: June 1st
STREET DATE: June 22nd
CAT. #3186
R/T: 90 minutes
*LIMITED GUARANTEE-DETAILS AVAILABLE FROM YOUR DISTRIBUTOR OR FROM MAGNUM ENTERTAINMENT

WINGNUT FILMS PRESENTS
PETE O'HERNE · MIKE MINETT · TERRY POTTER · CRAIG SMITH · PETER JACKSON · DOUG WREN · DEAN LAWRIE
in BAD TASTE
written and directed by PETER JACKSON · additional script TONY HILES, KEN HAMMON · music MICHELLE SCULLION
post-production supervisor JAMIE SELKIRK · sound mix BRENT BURGE · consultant producer TONY HILES · producer PETER JACKSON

MAGNUM

NINE

'ALL I WANT TO DO IS MAKE MOVIES'

*I*N 1988 I HAD THE JOB of introducing Peter Jackson to the world when I started selling his first feature film *Bad Taste* to overseas territories. *Bad Taste* was, after *Death Warmed Up*, New Zealand's second horror film. Peter had started making it five years earlier while working as a photo-engraver on Wellington's *Evening Post* newspaper. The film began as a short named *Roast Of The Day*. After a year of weekend shooting, and Peter's editing it at night on his parents' kitchen table, it had become an hour-long production titled *Giles' Big Day*.

Early in 1985 Peter had asked the Film Commission for $7,000 to complete the film. When he received Jim Booth's rejection letter expressing doubts about the film's commercial potential, he typed an eight-page reply saying he would complete it with or without commission finance. 'All I want to do is make movies,' he wrote. It wasn't Peter's first rejection: after leaving school he had tried without success to get a job at the National Film Unit.

A year later, Jim told me that word about the unknown film-maker was spreading in the industry. Peter was screening his incomplete film (which had cost him $15,000 up to that point) and the professionals were impressed.

Jim went to look at it, taking film-maker Tony Hiles with him. They came back with news of a maliciously delightful over-the-top alien splatter-horror comedy which they thought the commission should support. Not everyone agreed, but Jim persuaded the organisation to invest on condition that two experienced professionals were added to the team. Tony became consultant producer; he would also make a film about the making of *Bad Taste*. Jamie Selkirk was chosen as editor. Eighteen years later, by then one of Jackson's closest collaborators, Selkirk would win an Academy Award as editor of *The Return Of The King*, the final in *The Lord Of The Rings* trilogy.

In November 1986, when he received his first cheque from the Film Commission, 24-year-old Peter Jackson resigned from his job on *The Evening Post*. Within 15 years, without leaving home, he would become one of the world's top producer-directors.

Not only would he buy the National Film Unit, he would move the entire operation from Lower Hutt to the Wellington suburb of Miramar – not far from where it had begun its life in the 1940s. And he would replace the venerable Film Unit name with 'Park Road Post', a reference to the street address of his new lavish hi-tech post-production facility.

In mid 1987 Jim Booth and I viewed the new film-maker's work-in-progress and recommended the commission invest further money so it could be completed, with the proviso that the film was cut 'to overcome confusion and languor in the first half hour'. I then met Peter for the first time, to discuss how we would market his film. His personal credits on *Bad Taste* included not only producer and director but also camera operator, co-editor, makeup/special effects with Cameron Chittock, and actor (Derek and Robert). He designed the brochure cover as well.

We launched the film in the Cannes market in 1988 with a low-key, low-cost promotion, although there was nothing low-key about the image of the alien with the machine-gun which Peter provided for the brochure – and nothing low-key about the alien head, made of latex, which he delivered to the office, and which was worn every day by New Zealanders walking among the crowds on the Croisette. The synopsis was pretty good too – aliens from outer space seeking human flesh for an inter-galactic fast-food chain. Everyone paid attention when they heard that astonishing story line.

Peter and Tony Hiles arrived in Cannes before screenings began. It was their first visit. I told them I was ready to sign contracts with three distributors who wanted to buy the film on the basis of the alien image and the synopsis. Pre-sales were always hard to get and I was excited and pleased, given that I'd been less than certain about whether the film would succeed. But Peter didn't like the idea of his film being sold to people who hadn't seen it, so I postponed the pre-sales deals.

Although most of the bigger distributors proved to be nervous about the film's extremes of humour, gore and splatter, the market screenings were a success and there was plenty of enthusiasm. I negotiated sales to 15 countries, making sure that all buyers had seen the film before we signed their contracts.

With a budget of just over $295,000 – of which the commission had invested $207,000 – the film was theoretically in profit by the end of the market. Gross income from my Cannes sales would equal the budget figure. But before the investors could start recouping their money there were loans for marketing and completion costs to be repaid, plus the commission's modest 15 percent share of income to help cover its sales costs. In time, *Bad Taste* would earn more than three times its production budget. And like all the most successful films it would keep on earning, year after year.

One of my first sales was to Andre Koob, a leading distributor of genre movies in France. His publicity was boosted when *Bad Taste* was selected for the Paris International Festival of Fantasy and Science Fiction Films, where *Death Warmed Up* had won the grand prix four years earlier. Peter went to Paris for his first experience of international acclaim. He happily reported to newspapers at home 'a continual barrage of chanting and screaming, clapping and cheering' during the screening of his film, which won the festival's special jury prize.

Koob blamed himself when his 40-print theatrical release was a disappointment. He had chosen not to use the Jackson campaign showing the alien with the machine-gun. Instead, he had created an artful poster that showed someone floating in space. Only 876 people came to see the film in 15 Paris cinemas on the first day of its release. The first week's total attendance throughout France was only 15,000. But with fast growth in the home-video market, Koob was not out of pocket for long.

Peter's original *Bad Taste* campaign was used everywhere else, although in England the poster was banned from the London Underground until the position of the alien's uplifted finger was adjusted to protect the sensibilities of British commuters.

From the first screenings of *Bad Taste*, Peter was identified by the international film industry as an important talent. When we sent the film to the Edinburgh Film Festival, its official programme declared Jackson a director 'with more than a hint of inspiration and fearsome dedication'. The British Film Institute noted an emerging trend: with *Bad Taste* following on from *The Quiet Earth* and *The Navigator*, New Zealand had become a leading source of cinefantastique.

British admiration kept growing. When the film was ten years old, the BBC approached me to buy television rights. I knew Peter's films had fans inside the BBC, but never expected they would show *Bad Taste* to British viewers. In the event, the BBC became the first broadcaster in the world to show the film – although it was screened late at night, when mainstream viewers would have been safely asleep.

United States distributors hadn't come to see the film at its Cannes debut so I decided to offer them individual screenings. My first call in Los Angeles was at the fast-growing independent company New Line, which had earned big profits from the first four features of its *Nightmare On Elm Street* series and was preparing a fifth. I screened the film for Sara Risher, who was by then one of New Line's top executives and had worked on all the *Nightmare* films. When I told her about *Bad Taste* she said it sounded like the kind of film New Line would want to release. But after 20 minutes she stopped the screening, and said the film was not for her company. I had not expected to be so downcast when I left the screening room.

Twelve years later, when Sara and I were guests at a Los Angeles reception celebrating New Line's deal to finance *The Lord Of The Rings*, we recalled how the company had missed its earliest opportunity to become involved with Peter Jackson.

As it turned out, the response of every American theatrical distributor was the same. None was willing to release *Bad Taste* in cinemas: it was considered too extreme. So although we continued to place the film in festivals, its first American earnings came from a video release. The Magnum video company promoted the film with twelve more alien heads moulded in latex and hand-made by Peter and his colleagues in Wellington at a cost of NZ$50 each. Five years later I found a theatrical distributor in Las Vegas who was willing to seek cinema bookings, but income was minimal. We even tried pay-per-view television, which was then being trialled in a few cities. But *Bad Taste* wasn't what viewers wanted to pay for.

More than a decade later, when the film was acknowledged as one of the world's best-known cult movies, it would earn substantial revenue from the fast-growing United States DVD market. I licensed it to Anchor Bay, a Michigan-based company which was reissuing the work of key international directors, including Werner Herzog, a film festival favourite from the 1970s. The company flew the original *Bad Taste* negative and soundtracks to the United States, where it produced an impeccable new digital master for a collectors' edition which included Tony Hiles' documentary about how the film was made. The release was a success. The distributors' only disappointment was that Peter didn't have time to record a commentary for their DVD. By then he was working around the clock on *The Lord Of The Rings*.

Canadians had a more relaxed attitude to *Bad Taste* from the beginning, and a theatrical distributor signed a deal early on. However the cinema release was called off when the Ontario censor banned it 'for indignities to the human body'. In Australia, censors first refused to approve the film, then passed it with substantial cuts, and then negotiated fewer cuts for a more restricted certificate. Censorship was also an issue in Germany, where for ten years no one was willing to release the film. Then I met Oliver Krekel of Astro Films. He was the first German brave enough to agree to a deal, pay an advance, take delivery of the materials and challenge German censorship.

Much to my relief, censorship was not a problem in France, Italy, Belgium, the Netherlands, Japan, Mexico or Argentina. These were among the first territories to release *Bad Taste*.

Cannes in 1988 was notable for its extremes. As well as Peter Jackson's splatter comedy, I was launching a time-travel odyssey which had cost almost 20 times as much and was New Zealand's second entry in the Cannes competition.

Although Gilles Jacob had signalled his interest in Vincent Ward's *The Navigator* nine months before the festival, he hadn't confirmed his selection until four weeks before opening day. Nonetheless, the official competition screening became another memorable Cannes occasion. Again we assembled in Vincent's room at the Carlton Hotel. Again we walked down to limousines which drove us to the red carpet, where we walked up the stairs in front of the photographers to be welcomed by Jacob. After the screening there was a five-minute standing ovation, where the young director shared the spotlight with his leading actor, American Bruce Lyons. Phillip Adams in *The Australian* described the triumphal ending as 'one of the warmest responses I've heard in a lifetime at the festival'.

During our two weeks at Cannes we signed 16 deals for international theatrical releases of *The Navigator*, putting it neck and neck with *Bad Taste* as New Zealand's best seller. Associate producer Gary Hannam took no more than five minutes to sell the film to Chris Auty for British theatrical release by Recorded Releasing. Auty moved even more rapidly to license television rights to the BBC, perhaps because Alan Howden was standing next to him, ready to do the deal. A former film editor of *Time Out*, Auty was a long-time admirer of Vincent Ward's work. The connection was to prove durable: 16 years later he would be the British co-producer of Ward's fifth feature, *River Queen*.

British critics liked *The Navigator* but French critics were divided. One explained: 'The great film of this festival was vilified by the critics of the "great" press, spurned by the jury, but very popular with the public, who greeted it with enthusiastic applause.' Belgian critic Patrick Duynslaegher wrote: 'Scoffed at by many as being a moralising fable, *The Navigator* is primarily an ode to untrammelled imagination, a film in which everything is possible and the most incredible things can happen.'

The split among French critics meant that it took longer to find a distributor for France. I was on the verge of signing a deal with a small company for a small amount of money when I received a phone call from a high-profile member of the French cinema establishment, Anne Francois of film distributor AFMD. She was on her way to make an offer. Judy Russell's successor Rhys Kelly had to think of somewhere to relocate the other distributor while we met his more affluent competitor. She found an excuse to shunt him into the kitchen of our office-apartment, and closed the door just in time. After an hour Anne had agreed to sign a deal with a substantial advance.

Later I was told that the film's Canadian buyer Rene Malo would be paying for AFMD's French rights as well as his own. Six months after Cannes I flew to Montreal to finalise negotiations, which had stalled as we tried to merge the two complex agreements. After ten hours in the Malofilm board room, I was still arguing with the company's lawyer. I had to decide whether to accept an imperfect contract or let the negotiations collapse. A lot of money was at stake. I agreed on the contract and we signed it. The following morning I flew to Milan for the next Mifed market.

The Canadian release was a success, but the French failed because of the critics. There were only 21,000 admissions in the first four weeks. 'A poor result,' AFMD told us, with losses they would not be able to recoup from video or television sales. It wasn't until the arrival of *An Angel At My Table* two years later that French critics would accept a New Zealand film as worthy of praise.

The Navigator's official co-production status meant that, for the first and only time, we shared a Cannes reception with the Australians. The combined party on the beach at L'Ondine started at midnight after the competition screening. The Australians had been upset when the French described the film, because of the nationality of its director and its locations, as a New Zealand production. Nevertheless, it qualified for the Australian film awards, where it won best film, best director and four other categories. It also won 11 of the year's 12 New Zealand film awards, including best film and best director, and first prize at the Rome and Munich Fantastic Film Festivals, where it competed with *Beetlejuice* and *The Blob*.

Circle Releasing, headed by Ben Barenholtz, bought United States rights. When the film opened in New York in June 1989, Caryn James wrote in *The New York Times* that it was a 'dark thrilling fantasy that places Mr Ward, a 33-year-old New Zealander, among the most innovative and authoritative young film-makers'.

The American release was briefly challenged by the Disney organisation, which was nervous that the title might confuse audiences for its family film *Flight Of The Navigator*. Lawyers were hired and Disney wasn't satisfied until Ward's film had 'a medieval odyssey' added to its title in small print for the North American release.

The American poster, like others from the film's international releases, suggested the New York skyline rather than that of Auckland, the actual destination of the film's medieval travellers seeking to save their village from the plague by journeying to a city on the far side of the earth. For offshore movie publicists, New York was the only city which could be a movie destination.

Vincent paid attention to the smallest details of our Cannes campaign, just as he had paid attention to the smallest details of what had been an exhaustingly long and demanding shoot. He was particularly proud of the tiny captions for the photographs. He had designed the layout and typography with care, and checked that we had stuck all the captions on the backs of all the photos. When he was due to go to an official festival dinner, he at first refused to attend, insisting on staying in the office to affix some captions he discovered had not been attached.

Vincent later wrote a book about his experiences making *Vigil* and *The Navigator*. British director John Boorman, in the foreword, described the director's skill as consummate, 'but more importantly he is that rare monster, an artist with a remorseless vision that drives him into impossible places of folly and madness'.

Among other New Zealand titles at the 1988 Cannes market was Merata Mita's *Mauri*, a tale of an old woman (played by activist Eva Rickard), her granddaughter, and a man with a secret identity (played by Anzac Wallace of *Utu* fame). Mita's partner Geoff Murphy, who was the film's associate producer and also played the part of a racist farmer, said it was the first time the full Maori perspective had been seen on a cinema screen. *Mauri* went on to win a prize at Italy's Rimini Film Festival, but found few overseas distributors.

Nor did Melanie Read's second feature *Send A Gorilla*. Dorothee Pinfold had set up a new company, Pinflicks, and *Send A Gorilla* – a comedy about three young women trying to deal with mayhem at their singing telegram company – was its first production. Dorothee brought a gorilla suit to Cannes and the gorilla promenaded on the Croisette in competition with the *Bad Taste* alien. But the gorilla didn't attract as much interest.

There were two more Mirage productions at Cannes that year. Leon Narbey's first feature, *Illustrious Energy*, was a sad, beautifully filmed story of Chinese miners in nineteenth-century New Zealand goldfields. Leon had graduated from the School of Fine Arts at Auckland University and spent his early life as a sculptor and experimental film-maker before becoming one of New Zealand's leading cameramen. He had shot *Illustrious Energy* in the dramatic landscape of Central Otago, where he was visited by Pierre-Henri Deleau, who was on his third trip to New Zealand. Producer Chris Hampson drove Deleau to the locations on the sites of the original gold rushes, but the completed film wasn't offered a slot in the Directors Fortnight as we had hoped.

The other Mirage production, Larry Parr's directorial debut *A Soldier's Tale*, was based on a novel by M.K. Joseph about wartime loyalty and betrayal. Parr had long wanted to direct a feature, and *A Soldier's Tale* was the opportunity he created for himself. He also produced the film and shot it in France.

However, a saga was about to unfold. Mirage had become a public company in 1987, merging its interests with Don Reynolds' Cinepro production house to 'establish a quality international presence in film and entertainment markets'. At the Film Commission, Jim Booth believed the merger would provide more secure investment structures and enable a greater range of films to be made. He would be proved wrong. By 1988 Mirage was running out of money, in part because of a funding crisis that had hit the production of *A Soldier's Tale* when French investment failed to eventuate. By the end of our stay in Cannes we were distressed to learn that the company was in trouble.

Mirage was doubly disadvantaged because of the collapse of the American distribution company Atlantic Releasing, whose first New Zealand release had been *Smash Palace*. Atlantic had contracted to release *A Soldier's Tale* in the United States and to pay an advance of US$1 million. But the company had collapsed without releasing the film or paying the advance.

Early in June, Mirage was put into receivership by its major lenders, the Bank of New Zealand and Equiticorp, with debts initially reported as $4.5 million. The seven productions in the Mirage catalogue were taken over by receivers, who took their time making decisions; for six months the films were inaccessible, bringing protests from angry distributors who called us to complain that they had paid money but couldn't get prints for their releases.

Illustrious Energy continued its festival life, in spite of the uncertainty about sales. At first, because of a small unpaid debt, the National Film Unit was reluctant to provide a print. The commission paid the money so festival commitments could be met, and the film won a bronze charybdis at Taormina, where *Corriere Della Sera* praised its 'great formal rigour, vast horizons and profound lyricism'. It also won best film at the Hawai'i Film Festival, where the jury was headed by Richard Schickel of *Time* magazine. Its eight New Zealand film awards included best director.

There were several offers for the Mirage library. Although one was from a New Zealand company, the receivers sold all rights to a British investor who, we were told, offered the largest amount of money. Control of the films was moved to a sales company in Los Angeles.

Illustrious Energy languished in the company's big catalogue – even more when it was re-titled *Dreams Of Home*. On one occasion I was phoned by a buyer who had been told by the Americans that the film was not in their catalogue. Not surprisingly, therefore, it had few theatrical releases.

Sam Pillsbury's second feature *Starlight Hotel* had been produced for Mirage the previous year by Parr and Finola Dwyer. It had been launched and sold to 20 countries before the demise of the company. It earned some good reviews from United States critics, including *The Los Angeles Times* which called its story of two runaways during the Depression 'an enchanting odyssey'.

In the year of the Mirage collapse, Wellington-based producer Dave Gibson decided to close his sales company. With the disappearance of tax-sheltered investors, it was becoming harder to find films to sell and he decided to concentrate on the core business of production.

At the same time a third sales agency made a brief appearance. Energy Source International evolved from a production company which had been set up in 1980 to make Christian television programming. Englishman Peter Sainsbury moved to New Zealand to become executive producer and Australian Sue Thompson handled sales. But its life was short. Thompson resigned in May 1989 and two months later a receiver was appointed.

The Film Commission was once again the only local entity selling New Zealand movies. Some people argued that its activities had been unfair competition for independent sales companies. I never accepted this. The independents had always had the first choice of films to sell, and were always offered the chance to participate alongside the commission at international markets. Producers were free to choose who sold their films. I was happy that so many chose the commission.

Dave Gibson often stated that our 15 percent commission on sales was too low and couldn't be matched by independent sellers. A few years later, when Carole Myer was asked to sell two New Zealand films, we discovered that some offshore sales agents charged an even lower percentage, although their costs and overheads were inevitably higher than the commission's. So I stuck to my guns.

KITCHEN
SINK

A NIGHTMARE COME TRUE...

A FILM BY ALISON MACLEAN

TEN

HARD TIMES,
AND HARDER STILL

G EOFF MURPHY'S FIFTH FEATURE *Never Say Di*e was aimed at the American market. It had a budget ten times as large as that for *Goodbye Pork Pie* and starred a young Maori actor, Temuera Morrison, opposite Lisa Eilbacher, an American chosen because she had been in *Beverly Hills Cop* with Eddie Murphy. To add to the film's intended appeal to American audiences, George Wendt from *Cheers* had a cameo role, and there was plenty of money for stunts and explosions.

Produced by Murphy and Barrie Everard, *Never Say Die* premiered at Mifed in October 1988, and by that time most major territories except the United States had been pre-sold by the film's American sales agent, Kings Road Entertainment. But in spite of the American stars and the stunts and explosions and money, the film didn't get a theatrical release in the United States – unlike Geoff's three previous features, which had all found American distribution and success on the basis of their unadulterated and unapologetic New Zealand flavour.

Kings Road had, however, collected substantial payments from the film's non-American sales. When the producers completed delivery of the materials needed to supply these overseas buyers, a large share of the money was due to be sent to New Zealand. However the company kept delaying the payment, coming up with reasons which didn't seem at all reasonable. It was a tense time. The commission wasn't concerned only because it had invested half the cost of the film; it also wanted to ensure that earnings were received by the private New Zealand company that had invested the other half, a rare event at that time.

The delay led to an unprecedented move. The commission decided to hire a Los Angeles lawyer to deal with the problem of non-payment. Finance director Chris Prowse flew to Los Angeles to meet Tom Garvin, a contact from my first Milan market nine years earlier. Tom turned out to be a good choice as legal adviser. After he briefed Chris on the facts of the case, Chris went to meet Kings Road management, only to discover that the company had changed owners overnight. The legal case was, however, all on our side, and the overdue money finally arrived in the film's New Zealand bank account without the issue going to court.

A year later the situation was repeated. Kings Road was again being slow with payments, this time income from the film's international video sales. The commission hired Tom for a second time and got the same satisfactory result.

Meanwhile, Barrie Everard handled the New Zealand release with high expectations. He released 22 prints, and advertised widely. Despite this, the film attracted only 80,000 New Zealanders, an eighth of the number who had seen *Goodbye Pork Pie*. The unexpectedly low attendance didn't cover the release costs, but earnings from video and a substantial television sale made up the difference.

After the 1988 Mifed, I flew to Brazil for New Zealand's first official participation in the Rio Film Festival. Sue Thompson of Energy Source came with me in place of Paul Davis, with whom I had made an exploratory visit the year before, and who had moved to London after the collapse of Mirage. Rio had become the sixth biggest film festival in the world. Paul and I had discovered that – in spite of chaotic organisation – it offered a chance of meeting South American distributors, who were elusive at the other international events. Brazil was another territory where New Zealand, both the country and film industry, seemed to be unknown. On top of this, the organisers were offering to pay all our costs.

The festival was again chaotic, but in spite of our videotapes failing to play on any Brazilian video machine we managed to make some sales. I sold a package of features to a local television network and Sue sold her company's television series about the Greenpeace ship, the *Rainbow Warrior*. After hours, we were introduced to some of Rio's 24-hour festivities by New York-based Brazilian Fabiano Canosa, who had taken time off from his job as film programmer at the New York Public Theatre to look after the English-speaking guests.

Late one night Sue and I were mugged by three teenagers, who stopped us and held a broken bottle to my neck as we were walking to our Copacabana hotel after dinner. We had heeded warnings not to carry any valuables except for small change, and were trying to keep inside the brightly floodlit areas of the beachfront promenade. But the floodlights didn't cover every part of the street. After several stressful minutes we handed over our few cruzados, which I retrieved from my shoe. The robbers then threw away their bottle, ran to their car and drove away. My guidebook said muggings were a feature of life in Rio, and we noticed all the beachfront apartment buildings had armed guards.

The next night we dined in a restaurant recommended by the festival. We were the last to leave, after waiting for the bill which never came. The restaurant called a taxi, and everyone smiled as we walked out the door. But as we stepped into our taxi, our waiter came running out, apologising for forgetting us. We then discovered that the establishment didn't take credit cards. The policy was 'cash only', and after our experience with the muggers we weren't carrying much. What to do? I showed the staff our festival passes, and promised to return the next day with the money. Next evening I took a taxi back to the restaurant. The staff expressed shock that I had been honest enough to return, and refused to take the cash. They offered another free meal, which unfortunately I couldn't accept, being on my way to meet a potential buyer.

Ugo Sorrentino, one of the most successful art-film distributors in Rio, had invited us to a party at his home, where we admired his art works and noted that he seemed to have quite a few servants. The next morning I met him at his office, which was in a much poorer part of the city. He looked me in the eye and said, 'You realise I can't afford to pay you any money for this picture.' Negotiations had begun. *The Navigator*, which was in the Rio competition, became the first New Zealand feature released in Brazilian cinemas, with Ugo sending regular shares of the modest box-office takings.

1989 began with New Zealand's biggest-ever international retrospective – 78 features and shorts. Curated by Jonathan Dennis for the New Zealand Film Archive, it was held in the Italian city of Turin. *Te Ao Marama* was accompanied by a book about New Zealand cinema, edited by Jonathan and Italian film historian Sergio Toffetti in three languages: Italian, Maori and English. The book's overview of New Zealand culture went beyond cinema, with quotes from literary legends Lady Barker, Katherine Mansfield and Janet Frame, and reproductions of paintings by artists Ralph Hotere and Colin McCahon. 'After China, Vietnam and Black Africa, it is now the turn of one of the youngest and most promising areas, New Zealand,' Piemonte's cultural councillor Enrico Nerviani wrote.

A few months later New Zealand was back in the cultural embrace of France, with a short film in the Cannes competition for the first time. Alison Maclean's black and white *Kitchen Sink*, produced by Bridget Ikin, had wide appeal with its story of a woman who kept pulling a hair out of the plughole of her sink until she succeeded in extracting a sinister naked man. My colleague Rhys Kelly sold the film all over the world and negotiated selection in 18 more festivals.

Alison and Bridget joined us at Cannes, where we attended the competition screening of Jane Campion's Australian-made *Sweetie*, which was booed by some members of the audience.

Only one new feature was available for us to screen in the 1989 Cannes market – Richard Riddiford's *Zilch!*. Some people thought the title was a reference to the dearth of New Zealand films. Almost 70 features had been made since the commission was established, more than half with commission investment. But production was falling again. The commission's insistence on private investment supplementing its production funds, and longer periods of development intended to achieve better scripts, were having an unintended result: few features were starting production.

As a result, a mini-exodus of film-makers had begun.

Geoff Murphy accepted a job in Budapest directing *Red King, White Knight,* a telemovie for cable television company Home Box Office, starring Max von Sydow and Helen Mirren. Over the next 12 years, Geoff would direct ten American productions for cinema and television.

David Blyth went to Miami to direct the genre thriller *Red-Blooded American Girl*. John Laing went to Canada and France to direct episodes of a new series for HBO. Sam Pillsbury joined the expatriates in Los Angeles, where his first American feature, *Zandalee*, starred Nicholas Cage and was filmed in New Orleans.

From his new base in Sydney, Vincent Ward started developing *Map Of The Human Heart*, a four-nation (Australia, France, United Kingdom, Canada) co-production with Tim Bevan as British producer, Linda Beath as Canadian, and a role for Jeanne Moreau as a nun. The film's locations included the Arctic Circle. When it was released in 1992, the Cannes festival selected its third Ward film.

After this Vincent moved to Los Angeles, where he would remain until the end of the 1990s. He earned a lucrative writing credit for *Alien 3* and in 1998 directed his only Hollywood movie, *What Dreams May Come*, a fantasy about life and love after death, starring Robin Williams, Cuba Gooding and Max von Sydow.

It seemed the only director who had no intention of leaving home was Peter Jackson.

Jim Booth, whom Jackson described affectionately as a kind of bureaucratic pirate, had left the Film Commission at the end of 1988 after six years as chief executive. He joined Peter Jackson as producer at WingNut Films.

The company's first film was to be *Braindead*, another horror movie, this time about a young man's mother who is bitten by a monkey and becomes an insatiable man-eating zombie. However when he and Peter found it hard to raise enough money and a Spanish investor withdrew, Jim persuaded the commission to instead support a lower-budget project called *Meet The Feebles*.

Jim and Peter came to the first board meeting of 1989 – which was also the first meeting attended by Jim's successor – and brought some of the 'real-life' puppet characters from the film they wanted to make. Peter said he felt annoyed that puppets were shown only as cute and innocent. His puppets would have mental problems, drug addictions and sexual obsessions. Board members' reservations about a porn movie sub-plot were overcome by the energy of the presentation, plus their recognition of the growing success of *Bad Taste* and the talent of its young director.

The 'spluppet' comedy was ready for launching at Mifed later that year. Peter arrived towards the end of the market, exhausted by completing his film to a tight deadline and with an even tighter budget, which had been topped up by the commission – after tense and difficult negotiations – several times. He spent most of his time in the office of his new sales company Perfect Features, which had been chosen by Jim after meeting its founder, Grace Carley, at Cannes the year before.

The market screenings of *Meet The Feebles* attracted full houses, with much laughter and applause. The commission committed extra money to blow up the film to 35 millimetre from the 16-millimetre gauge on which it had been shot, and Carley finalised plenty of sales. But not everyone could cope. A BBC buyer told me that although he was impressed by *The Deer Hunter* parody (in which a frog suffers from extreme post-Vietnam shell-shock), he could never show the film. The United States was again slow to acknowledge the Jackson style. *Meet The Feebles* would not get an American release until the mid 1990s, when *The New York Times* commented: 'It's possible to admire Mr Jackson's irrepressible fancifulness without remotely admiring the uses to which it has been put … sophomoric sight gags, limiting his audience to only the most ardent fans of knee-jerk decadence.'

Peter had flown to Milan from London with his partner and co-writer Fran Walsh. We had issues to talk about. The British distributor of *Bad Taste* had failed to complete payment of the advance. Peter was not happy about the breach of contract, but we thought we could work out ways of fixing everything.

First, we found a theatrical distributor, Joe D'Morais of Blue Dolphin Films, who was willing and able to pay the remainder of the advance. The contracted distributor then allowed Joe to organise a theatrical release, which opened at the big Prince Charles Cinema off Leicester Square. Our share of the theatrical income was to be channelled through the distributor with whom we had signed the original contract. But he again failed to send our share of the money.

In response to pressure from the commission and Peter, he mailed a cheque. It bounced. A month later he faxed: 'I am at present out with the police busting a piracy racket which I regret to tell you includes *Bad Taste*. Police expect an arrest imminently.' It sounded like a scam.

A year later we discovered that pirated copies of *Bad Taste* were on sale in London video stores, although by then we had terminated the deal. We tracked down a company that claimed to have obtained the rights from our original distributor, who then phoned me to claim he was 'bankrupt, unemployed and destitute' because of a 'terrorist-run, IRA-funding piracy ring'. If Peter hadn't been so angry, the convoluted claims could have been considered for a movie script.

It took three years to end the piracy in Britain. By then, Peter's reputation had grown to the point where a substantial company was willing to license *Bad Taste*. I signed a deal with Polygram and from then on there were regular reports and payments from legitimate sales of *Bad Taste* videos in the United Kingdom.

But most of these problems were still in the future when I left the 1989 Mifed market and flew to Germany, where a New Zealand film retrospective was to be launched by the Hamburg Film Institute, with John O'Shea and Wi Kuki Kaa as honoured guests. The New Zealand ambassador brought New Zealand wine in the boot of the embassy car when we turned up for the opening night reception. It would be the only time we gave a party and fewer than half the guests arrived. The others had stayed at home to watch historic events unfolding on television. That night the Berlin Wall came down.

The next day the retrospective proceeded as scheduled: one night after the beginning of the end of the world's 40-year Cold War, *Goodbye Pork Pie* screened in Hamburg to a sold-out house.

Jim Booth's successor as head of the Film Commission was Judith McCann. Born in Christchurch, Judith had lived in Ngaruawahia and Tauranga until her parents moved to Canada, where she graduated in history from the University of Saskatchewan. She had become the Canadian government's chief of film policy and certification before joining the Canadian Film Development Corporation (later re-named Telefilm Canada) and becoming its deputy director. When Judith had visited New Zealand in 1987 to negotiate a government-to-government co-production agreement between Canada and New Zealand, it was the first time she had been back to her homeland since she was a schoolgirl.

In Canada, Judith had learnt French as part of the government's policy of bi-lingualism. Given this experience, she was at pains to ensure that New Zealand's official second language, Maori, was one of the languages in the agreement. When the documents were later signed at the Vancouver Film Festival, Barry Barclay and Wi Kuki Kaa (who were festival guests for screenings of *Ngati*) raised the issue of why Canada's aboriginal languages were not also included. They were added when the agreement was reviewed three years later, with Judith this time representing New Zealand instead of Canada.

Judith started work as executive director of the Film Commission in January 1989. In one of her first moves, she decided staff should become fluent in Maori. Weekly lessons began. We were introduced to Maori protocol and found ourselves learning waiata, which we sang at official events. The commission would later appoint six kaumatua and kuia, respected Maori elders, to provide advice from a Maori perspective. One was 1940s movie star Ramai Hayward.

All, though, was not well. The new chief executive was facing a slump in film production – the second in five years. After Jim Booth's departure from the commission, he had, as a newly minted producer, talked about 'a terrible crisis of confidence in the film industry'. The environment for producers, he discovered, was 'very hard, and it has been for ages'.

McCann persuaded the commission to deal with the situation by changing its policies. Instead of holding out for the elusive partial investment from private sources, board members agreed that to get the industry working again the commission would invest 100 percent of the cost of five new productions. The policy succeeded. There was a resurgence of production. And one of the five films would become a New Zealand icon.

EIN ENGEL
AN MEINER TAFEL

An Angel At My Table

Ein Film von Jane Campion

Großer Spezialpreis der Jury   Intern. Filmfestspiele Venedig '90

ELEVEN

THE EARTH MOVES IN
NEW YORK – AND LOS ANGELES

*I*N A BOX ABOVE THE AUDIENCE at the New York Film Festival, three
New Zealand women stood together in the spotlight. It was September 1990,
and the full house of two thousand people in Lincoln Centre's Alice Tully
Hall was giving a standing ovation to movie director Jane Campion, actor Kerry
Fox and producer Bridget Ikin. Their film, *An Angel At My Table*, had been the first
New Zealand feature ever selected by the New York festival. The screening, with the
enthusiastic reaction of the audience, was one of my happiest moments in the film
business. It was also the end of a prolonged period of dispute and disagreement.

Three years before, Bridget and Jane had brought the Film Commission their
proposal to make a television mini-series of *To The Is-land*, the three-volume
autobiography by novelist Janet Frame. The commission didn't want to finance a
television series. But it did want to support a New Zealand project by Jane Campion.
The daughter of Richard and Edith Campion, founders of the legendary New
Zealand Players theatre company of the 1950s, Jane was living in Sydney, where
she had studied at the Australian Film and Television School. In Australia she had
directed three short films, a tele-feature, and *Sweetie*, her first feature, produced by
John Maynard, who now had an office in Sydney as well as Auckland. All her films –
short *and* long – had earned official selection at Cannes, an extraordinary recognition
of her abilities.

Jane's desire to direct a New Zealand film presented another chance for the
commission to enable a talented expatriate to return home. And the story of how
Janet Frame overcame personal disasters and illness to become New Zealand's most
celebrated author was one that seemed ideally matched to the director's talent.

Although the project was to be shot on locations in Spain, France and England
as well as New Zealand, it was to be made for only NZ$2.6 million. The commission
agreed to commit the total budget. Unlike most of its investments, this one didn't
leave it totally at risk: the costs would be partly underwritten by Channel 4 in England
and the Australian Broadcasting Corporation, both of whom agreed to buy the

mini-series when it was completed. However they would not pay until the production was delivered.

To satisfy the commission's policy of supporting productions for cinema rather than television, Jane and Bridget agreed that in New Zealand the mini-series could be screened in cinemas before it was telecast. All of us at the commission hoped Jane and Bridget would change their minds and agree to theatrical release all over the world, but we didn't force the issue because they were so strongly opposed.

I first discussed the project with buyers at the 1989 Cannes Film Festival where *Sweetie,* while not popular with some of the festival audience, was widely admired by the industry. By the time I arrived at the Mifed market six months later, Jane's reputation was growing and art-film distributors from all over the world were asking how they could acquire rights to her new work.

By February 1990 she had completed the first two episodes and the production had been renamed *An Angel At My Table*, the title of the second book in the trilogy. The commission had the rights to sell the series for release on North American video and television, so I took videotapes of the two episodes to the American Film Market in Los Angeles.

I showed the tapes to only a few distributors. One of them was Ben Barenholtz of Circle Releasing, who had released *The Navigator* two years earlier. As soon as Barenholtz viewed the two episodes he said to me, 'This is too good to be wasted on television.' He was quoted in *The Hollywood Reporter* as calling it a masterly work. Other distributors confirmed his view. I received offers of theatrical deals from distributors in Italy, France and Holland, and returned from Los Angeles convinced of the theatrical potential of the production.

Jane and Bridget, however, were resolute. They would not agree to a cinema release outside New Zealand, and wouldn't allow me to accept any of the offers. I had never faced such a frustrating situation.

Ben and I worked out a convoluted contract by which Circle Releasing would acquire United States video and television rights, with the right of first refusal to release the film in cinemas if theatrical rights became available. My pressure to make this happen made me more unpopular with the film-makers. Buyers were angry too. They couldn't understand why theatrical screenings would not be allowed. I told them the director was undecided about a theatrical release outside New Zealand, although I couldn't explain why. I really didn't understand, but I wasn't going to give up. The stand-off continued as Jane completed work on the third and final episode.

Four months after the Los Angeles market, I arrived at the 1990 Cannes Film Festival with tapes of all three episodes in my suitcase. They weren't for general viewing, but I showed them behind closed doors to some of the distributors who wanted to release them in cinemas. They all repeated their belief that the production should be available for theatrical audiences.

Two days before the end of the festival I faxed a long letter to Jane, telling her of the buyers' enthusiasm and my frustration. Two days later I received a message hinting that Jane might change her mind. Four days later I was in our Cannes office late at night, working through my notes and writing scores of letters. It was embarrassing to be writing to so many people telling them they couldn't buy something about which they were so excited.

I was hoping to hear from Jane. But the phone didn't ring and the fax machine was silent. Some time after midnight, I decided to phone her to see if she would be willing to talk. My timing was right. We spent an hour reviewing the different points of view. I talked fervently about the passion of the theatrical distributors who wanted to show her work. The enthusiasm finally got through to her. She agreed to allow me to start selling *An Angel At My Table* for cinema release.

I stayed in the Cannes office for four more days, faxing distributors and starting to negotiate theatrical contracts.

Plans had already been made to test the three episodes a few weeks later at the Sydney and Melbourne Film Festivals. The response from Australian audiences and critics was outstanding. The Sydney audience voted them collectively the festival's most popular film and gave them a standing ovation. The following month the three episodes had a New Zealand premiere in the Wellington and Auckland Film Festivals, where the response was the same. 'This is television work of a high order,' wrote Bill Gosden in the programme. 'We are delighted to present these preview screenings on the bigger but not huge screen which they deserve.' He noted that Janet Frame had called the work 'delightful'.

By the time of the New Zealand film festivals, Jane was working on the theatrical version. The Film Commission had provided money for the episodes to be converted from their original 16-millimetre gauge on to 35 millimetre as a feature film. Extra stress was added to the 'blow-up' process when the negative of the first episode was damaged while the first prints were being made. Three ways of repairing the damage were discussed, and the problem was overcome after much tension.

The theatrical version was six minutes shorter than the total running time of the three episodes. Jane had decided to remove a few scenes, but most of the difference was because two sets of opening and closing credits had been eliminated.

One of the many people who had hoped the television series would become a feature film was Guglielmo Biraghi, who was now director of the Venice Film Festival. He was quick to select the new feature for competition at Venice, the first New Zealand film to be so honoured. I flew to Venice to handle sales. The situation was complicated, not only because there were now two versions but also because sales rights were shared between two organisations. The deal with Channel 4 in London had given it the right to license television rights throughout Europe, while I controlled theatrical and video rights in the same countries. But there were no problems: everyone wanted to work together to ensure the best result.

Before the festival began in Venice, I licensed Italian theatrical rights to Roberto Cicutto and Luigi Musini of Mikado Films, who had already bought the television rights from Channel 4. They had made their name with successful releases of films from top British and French directors, but this was their first New Zealand film.

Roberto and Luigi were generous hosts. To handle the media, they invited the multilingual film publicist Simona Benzakein to join us from Paris. Jane was staying at the Hotel Des Bains, where Visconti had shot part of *Death In Venice*. Roberto and Simona became good friends of Jane's, helping her through the stress of becoming one of the most popular directors at one of the world's top three film festivals – the opposite of her Cannes experience the year before.

When we entered the Sala Grande for the official Venice screening, which began at 10.45 at night, Jane received a standing ovation from the thousand-strong audience. When the film finished at 1.20 a.m., she again received a standing ovation. It went on for five minutes. Then she was mobbed by friendly crowds outside the theatre.

Earlier in the day there'd been also been a standing ovation for Jane from the journalists at a crowded press conference. We were told this had never happened before, in the 40 years of the film festival. We counted positive reviews in 16 Italian newspapers. In *La Repubblica*, Irene Bignardi wrote that Jane was a precocious and poetic talent. *Il Manifesto* said the Venice festival had been 'a triumph for the New Zealander Jane Campion'. In Paris, *Liberation*'s admiring review of the film covered two pages. At last French critics were praising a film from New Zealand.

We started to get hints that *An Angel At My Table* was in line for one of the festival's main awards, so we decided to surreptitiously change Jane's travel bookings, and then invent excuses to ensure she stayed in Venice until the end of the festival. On closing night, the jury chaired by Gore Vidal (with Gilles Jacob as one of the members) awarded the film the special jury prize. It then won seven other Venice awards, and a British newspaper headlined its report 'Campion wins Venice'. After the awards had been presented, we were taken in a gondola to a gala dinner. Martin Scorsese, whose *Goodfellas* had won the festival's best director prize, shook Jane's hand and congratulated her.

One of my challenges in launching *An Angel At My Table* had been to reposition the director as a New Zealander, because her previous films had been made in Australia. This, I said repeatedly, was a New Zealand film by a New Zealand film-maker. The strategy worked. Jane became known as a New Zealand director, albeit one who lived offshore.

An Angel At My Table went on to win awards everywhere – the international critics' award at the Toronto Film Festival, the Otto Dibelius Award at the Berlin Film Festival, best foreign film of the year in the United States and Australia, and best film of the year in Belgium (where it beat *The Silence Of The Lambs* and *Barton Fink*). Its success helped bring Frame's autobiography to a wider audience, with the first foreign-language editions appearing in Italy, Germany and Scandinavia. It also spurred the first North American paperback edition – although the author's long-time but frugal New York publisher George Braziller at first seemed reluctant to commit money to it.

Within a year I had sold Jane Campion's film to almost all European territories. For the German sale I was approached by Karl Baumgartner of Pandora, who had handled the release of *Sweetie*. A French sale came quickly and the film had a two-year run in a Paris cinema. It was England's top-grossing film for its first two weeks. But the United States deal with Circle Releasing never took effect. Less than a year after I had first shown two episodes to Ben Barenholtz, he told me he was planning to leave Circle, and without him the company would not go ahead with the release we had been planning. After a month of correspondence we negotiated a new American agreement, this time with Fine Line, a newly established speciality division of New Line Cinema, represented by Tony Safford who had joined the company from the

Sundance Institute, where he had been working with Robert Redford. This was the start of New Line's relationship with New Zealand films.

By the time of its United States release in May 1991, the film's reputation had been well established by reviews from the New York Film Festival. The day the film screened at the festival had been memorable for many reasons. Before we went to Lincoln Centre there had been a small informal reception attended by the veteran publicist Renee Furst, who had been hired by New Line to promote the American release. Renee worked on all the top art-film releases; in her big midtown Manhattan apartment you could always expect to meet famous film-makers. She had been involved with New Zealand releases since *Sleeping Dogs* and *Smash Palace*.

We hadn't seen Renee for weeks, and when she arrived she drew attention to her slimness, which was noticeable as she had always been a large woman. 'My diet really worked,' she said, waving a walking stick. Her bravura was her way of covering the truth. She had been in hospital for cancer treatment, and would die within weeks. But she didn't tell us, and so we were unable to give her any sympathy.

After the screening and the wild, intoxicating applause, Jane was taken off to be fêted. Bridget and I walked to the Plaza Hotel, where we sat and talked until the bar closed at 3 a.m. Our long and sometimes angry dispute was over.

It wasn't until 2001, when I was leaving the commission and Jane spoke at my farewell, that I discovered why she had been so opposed to a theatrical release. She explained that the experience of premiering *Sweetie* at Cannes had been painful. The huge audience had been divided, and their reaction had included shouting and booing. Because of this, she had decided she didn't want to get into another situation where a theatrical audience could respond negatively to one of her films.

She had also worried that she might be judged badly because *An Angel At My Table* would be seen as 'tame' in comparison with her first feature. And of course *Angel* had been structured for television rather than cinemas – something which, surprisingly, had not proved to be an issue for anyone.

'Many people have helped me and been kind to me in my career, but no one has helped me so much under so much protest,' she said. Until that night, I had never understood the depth of the concerns she had felt in those fraught days 12 years before.

As things turned out, she need not have worried. In *The Christian Science Monitor* David Sterritt praised *An Angel At My Table* as superbly crafted and endlessly involving. In American *Vogue,* author Phillip Lopate, who had been on the New York selection committee, wrote perceptively: 'By harnessing her talents to someone else's story and overcoming her earlier fear of sentimentality, Campion has helped herself to mature as an artist. If *Sweetie* was Campion's dark angel, then Janet Frame is her light one, equally necessary in the long run.'

An Angel At My Table was a resounding success in cinemas everywhere. In New Zealand it ran for eight months, attracted more than 100,000 people and won six awards, including best film, best director and best actress. In Australia it ran for seven months and grossed three times as much as *Sweetie*. In Britain it ran for eight months, including four in London, and in Italy for over a year.

Jane visited many countries to publicise her film. She was sometimes reluctant to travel and anxious about meeting the press, but once she had made the decision (sometimes after at first postponing, causing anxious moments for journalists with deadlines) she charmed interviewers with her openness and frankness in answering their questions.

The level of interest in her work was extraordinary. In Berlin she had three days of meetings with German journalists, including six television interviews. Her final trip was to Japan. I had sold the film to the Shibata Organisation, run by the daughter of Japan's most distinguished motion-picture family, from whom I had bought Japanese features for film societies. A hundred journalists attended Jane's Tokyo press conference. They asked how all the key people who worked on the film could have been women.

The success of *An Angel At My Table* brought unexpected contact with Janet Frame. The French distributor Leonardo de la Fuente had been moved by a scene in the film which showed that when the then-unknown author first visited Paris she could only afford to stay in an attic. He told me he wanted to offer her a first-class ticket to Paris, with accommodation in a five-star hotel.

Bridget Ikin gave me Frame's unlisted phone number and I rang to tell her about the offer. At first she was interested, and started to list a number of cities where she had friends whom she would visit as part of her round-the-world trip. However within a few weeks her interest had waned. She said she had been happy in her attic, and she preferred to stay at home and start work on a new novel. I gave her my phone number in case she changed her mind.

Soon afterwards, Mikado asked if we would arrange for the influential daily *La Stampa* to interview her. Frame didn't want to talk to an Italian journalist, but she agreed to answer questions if they were submitted by fax and to give me the answers over the phone.

'I realise the immense choice that is made in creating an autobiography – not only the material but also the method of recording it,' she said. 'The film's scriptwriter and director were faced with a similar choice – to choose from so much material. I limited my life with my choice of words; the scriptwriter had to choose again, with further limits. When I read the beautiful script, I remembered how a proofreader sometimes changes words and that changes the meaning … With the magic of printing, such changes become fixed.

'It's the same in a film. A film is a new creation. It reminds me of a poem I once wrote about a turkey. It appeared in print as being about a turnkey. People thought I was writing about a prison. Of course, a turkey does live in a sort of prison …'

Frame said that her three books could have been about any family in New Zealand at that time. 'People of my generation all recognise the stories. New Zealand was then a country of sameness. It's now a country of differences, yet language has changed into a sameness.'

My note-taking couldn't keep up with the crisp flow of precise thoughts and ideas. She said that when she had spent a week on location, she had been amazed by the teamwork of the film crew. 'I was full of admiration for all the technical skills, even the person doing the time-tabling. I enjoyed the language of film – words such as "rushes". But a film first celebrates the image, and only then the language.'

About a year later my phone rang. It was Janet Frame again. She wanted to ask about her share of the film's profits. She had read about the film's international successes, and asked why the cheques she was receiving were for only small amounts. I explained that she was not receiving a share of the international box office. Her agreement provided for her to be paid a small share of the producer's share of the net profit, a figure which was calculated after the deduction of all production and marketing costs. She understood, and we parted amicably.

An Angel At My Table overshadowed the achievements of the year's other New Zealand films. There were standing ovations when Gaylene Preston's second feature *Ruby And Rata* had its world premiere at the Wellington and Auckland Film Festivals. The gentle comedy about a small boy and two determined and manipulative women stayed in New Zealand release for more than six months and was seen by more than

67,000 people; it was named on three 'ten best' lists and won four New Zealand film awards. It was voted one of the most popular titles at the Sydney and Melbourne Film Festivals, and won selection for the Montreal, Toronto, Los Angeles, Hawai'i, London and Munich Film Festivals.

Gaylene took *Ruby And Rata* to the Giffoni Film Festival in Italy. Travelling with her four-year-old daughter Chelsie, she arrived at Naples airport after a 40-hour trip from New Zealand to discover the festival had forgotten to meet her. She hired a taxi and was driven through the countryside to the village of Giffoni – where she faced an argument with the festival organisers when they refused to pay the $100 fare. Bad feelings were forgotten when the film won a citation for its originality, humour and stylistic sophistication.

Peter Jackson's *Meet The Feebles* was also selected for an Italian event – the Rome Fantastic Film Festival – where it won three prizes, including best director for Peter and best actress for Heidi the hippopotamus. In France the film won Le Prix Très Special, awarded by 13 Paris critics to the most 'out of the ordinary' film of 22 not yet released in France. Even with the award, however, it took two more years before a French distributor was found.

In Australia the censors approved the film with an unrestricted certificate, although, as Peter told Wellington's *Dominion* newspaper, it was in many ways more subversive than *Bad Taste,* which had been banned in Queensland for three weeks until the Board of Review was sacked by the state premier.

Martyn Sanderson's *Flying Fox In A Freedom Tree*, filmed in Samoa, was based on a novel by Albert Wendt. When it premiered at the 1990 New Zealand film festivals, *New Zealand Herald* critic Peter Calder commented cryptically: 'The visual approach compares well to the head-cleansing clarity of walking in a rain forest after a downpour.' The film had a United States premiere at the Los Angeles Film Festival, and a European premiere at France's Amiens Film Festival, where it was given a prize for its contribution to understanding the culture and identity of a people.

It was selected for competition at the Tokyo Film Festival, giving me a reason to fly to Japan for my third visit, both to represent the film and to meet the distributors with whom we were now doing business. By now more than a dozen New Zealand films had been released in Japan, although mainly on video. The highest price had been earned by *Shaker Run*, which had been sold by Mirage to a company with the distinctive name of Joy Pack. *Flying Fox In A Freedom Tree* won the best screenplay award for Martyn Sanderson but this recognition didn't secure a Japanese sale.

Producer Grahame McLean handled the New Zealand release, which began in Auckland's Queen Street and targeted movie-goers from the city's large Samoan community. Attendances were small. The film closed after a week. Samoans go to the movies, Grahame told me later, but not to *this* movie.

Rhys Kelly and I were now committed to three film markets a year, each with a different routine. It was February when we set up the first New Zealand office at the 1990 American Film Market in Los Angeles. We had been encouraged to attend by John Barnett, who said the commission should be seen there. The market was held in the old Beverly Hilton Hotel, a building familiar from the opening titles of the *77 Sunset Strip* television series, which had been popular in the early days of New Zealand television. Los Angeles was hit by a strong earthquake on the second day of our visit, which enabled me to tell a buyer that the earth moved when I met her.

Routines at the three markets varied greatly. In Los Angeles we rented a cheap car to drive to the market from the hotel where we were staying, and drove back each evening in the dark of the late Californian winter. At Cannes, which took place in the early summer, there were no transport costs – we walked to and from the office, and indeed we walked (and sometimes ran) all over town to screenings and meetings. Mifed in Milan was an autumn event, involving a subway ride from our hotel to our stand in the Fieramilano, and a subway ride back again after dark. We learned to beware of pickpockets after my wallet was stolen while I was travelling back to our hotel on the last night of a particularly tiring market. We also learned it was necessary to book in advance when we wanted to take a night off and attend a performance at La Scala.

After ten years of markets and campaigns and festivals and awards and premieres and releases and reviews, New Zealand was no longer unknown. Scores of regular buyers came to see us at each market, checking the new titles and wanting to discover the next big name. They all asked the same question: how does a country with a small population produce so many film-makers of international standard? I would answer by telling them that New Zealand film-makers had the freedom to choose the stories they wanted to tell – something Peter Jackson would talk about in an interview with *The New York Times* when *Heavenly Creatures* was released: 'I have a freedom in

New Zealand that's incredible. I can dream up a project, develop it, make it, control it, release it.'

We took Gregor Nicholas's first feature film *User Friendly* for its first trade screenings at our first American Film Market. We also took four small wooden dogs, look-alikes to match the film's dog-goddess who supposedly contains the elixir of life. Michael Kutza, director of the Chicago Film Festival where three of Gregor's short films had won prizes, selected the feature for his competition. He asked that we send the dogs as well.

I sold the eccentric comedy to Nikkatsu, a Japanese distributor who specialised in an unorthodox mix of children's films and soft-core porn, and to a German distributor who felt the dog theme offered marketing opportunities. But otherwise the film was not widely seen. Audiences didn't warm to what *The Dominion*'s critic Costa Botes described as 'perverse sexual humour … a rarity in the generally Calvinist run of New Zealand films'. Gregor's career would get back on track with his next film, a short titled *Avondale Dogs*, which won a dozen international awards.

User Friendly producer Trevor Haysom didn't produce another feature for 14 years, when his *In My Father's Den* became a substantial success in its New Zealand release. His fellow producer Frank Stark never produced another. Later he would become the second director of the New Zealand Film Archive and mastermind its expansion into spacious new premises in central Wellington.

It was always disappointing when, despite New Zealand's growing impact internationally, some films did not sell. In February 1991 a group of commission staff flew to Gore for the premiere of John Day's erotic thriller *The Returning*, which had been filmed in an old farmhouse near the small southern town. This was another feature which fell short of expectations. Although *Metro* reviewer John Parker said the film's story of a man who falls in love with a ghost was impressive as well as erotic, international sales were hard to come by – perhaps because of the film's leisurely pace. A year later John re-cut it to create a speedier opening. But distributors never give films a second chance.

Barry Barclay's second feature *Te Rua* was also little seen. Made partly on location in Berlin, with investment from two German film funds as well as the Film Commission, its story of young Maori people struggling to get back stolen tribal carvings from a European museum carried a vigorous anti-colonial message. Perhaps it was too vigorous for German audiences. I had pre-sold it to Wolfram Tichy's new German distribution company, but after he viewed the completed film he refused to release it. The German film festivals, usually enthusiastic about New Zealand cinema, also refused to show it. They would not explain their rejection. The film was never screened anywhere in Germany.

Its world premiere was at the 1991 Wellington Film Festival, after which it went to the Montreal, Toronto, London and Hawai'i Film Festivals. Peter Calder, while praising the film as an awesomely impressive achievement, found it 'almost brutally uncompromising to Pakeha eyes' – a clue to the negative German reaction and small local attendances.

Te Rua was the ninth and last feature produced by John O'Shea, who had been making New Zealand films for 40 years and had worked with Barry Barclay for 20. It would be ten years before Barry made another feature – a dramatised documentary about New Zealand's earliest settlers, the Moriori, which he called *The Feathers of Peace.*

1991 also saw the advent of Ian Mune's long-cherished film *The End Of The Golden Weather.* Mune had been developing the film since 1976, working with playwright Bruce Mason to adapt Mason's one-man play in which an eleven-year-old boy learns the difference between fantasy and reality during a 1930s summer holiday.

The play had become a classic of New Zealand theatre. But when Mason died in 1982 Mune seemed to lose confidence in the project. After the popular success of *Came A Hot Friday,* he made a second feature for producer Larry Parr, an action-adventure titled *Bridge To Nowhere.* Three years later he directed *The Grasscutter,* a thriller for television produced in Auckland by Tom Finlayson for the United Kingdom's ITV. Aired at the end of the 1980s, it attracted a British audience of more than 12 million, the biggest audience of Mune's career.

Mune's confidence in his *Golden Weather* project had been rekindled when he saw the similarly themed 1985 Swedish feature *My Life As A Dog.* With Christina Milligan as producer and Don Reynolds as executive producer, he re-worked the script, then successfully sought finance from the Film Commission and Television New Zealand.

We chose the Toronto festival for the world premiere. Standing at the back of the cinema during the first screening, Don Reynolds and I felt the Canadian audience's enthusiasm dissipate at the end of the film, when the endearing but tragic character

Firpo failed to win his race. The anticlimactic ending may also have been a factor in the film's New Zealand release. It ran for nine months, but its local theatrical audience of 68,000 was less than a fifth the number who had seen *Came A Hot Friday*. Nevertheless, reviewers were enthusiastic and the film won eight New Zealand film awards, including best picture and best director.

The film earned recognition in the United States when it was voted outstanding foreign film at the Youth In Film awards in Los Angeles. Ironically *My Life As A Dog* had won the same prize seven years earlier and had gone on to success in American cinemas. But *The End of the Golden Weather* didn't have the same appeal for US distributors, and never had an American release.

In England, the film gained a theatrical release – and a royal premiere for New Zealand at last. Princess Anne and her new husband Timothy Laurence attended the London occasion, greeted by Mune and Reynolds, as well as Dr Diana Mason, the widow of the playwright. The New Zealand Press Association's report was headlined 'Accolades for NZ movie' but after the film was released the agency reported unkindly: 'NZ film flops in London'. *Sight and Sound* found the film 'a little too pat to be poignant'.

In Germany it was re-titled *The End Of The Golden Summer*. Its cinema release was delayed for a year when there were cuts to television budgets: the distributor needed the television licence fee to pay for the German language version.

Financial problems were also an issue when David Blyth, who had been living in Los Angeles, returned home to shoot a family fantasy film titled *Moonrise*. Michael Heath's fanciful script was based on one of his radio plays. When the production lost its private investment, producers Murray Newey and Judith Trye decided to press ahead with a much-reduced budget. They removed elaborate flying effects and saved a million dollars.

For the leading role of the 600-year-old grandfather-come-friendly-vampire, Blyth cast 81-year-old American television star Al Lewis from *The Munsters*. 'If Walt Disney wasn't dead he would buy this movie,' Lewis told me when I met him outside the restaurant he ran in New York's Greenwich Village. His name helped me negotiate a video deal in the US, although with Republic Pictures and not the Disney empire. The company renamed the film *My Grandpa Is A Vampire* and put the star's face and name on the front of the video jacket. It earned big American sales.

This was a time when we liked the idea of New Zealand films having their world premieres in other countries: audiences at home could be buoyed by the perception of success and acclaim offshore. *Moonrise* had its world premiere at the Brussels Festival of Science Fiction and Fantasy Films, which also presented a Blyth retrospective with David introducing his earlier films.

But the film's theatrical release in New Zealand got off to a bad start when it was scheduled only at school holiday matinees, against tough competition from high-profile Hollywood titles. We failed to persuade distributor John Barnett to give the film a chance at evening sessions – it didn't seem possible for him to persuade cinemas to book the film at night. For whatever reason, Hollywood competition or local inertia, the release ended dismayingly, with less than 2000 attendances. It was the smallest local audience for any film with commission investment. I thought the film deserved a better result.

Ever optimistic, I had high hopes when producer Don Reynolds asked me to handle sales of *Old Scores*, a comedy about rugby football – a new genre, perhaps. The co-venture between New Zealand and Wales was about two teams who replay a test 25 years on. It starred former All Blacks Grizz Wylie, Waka Nathan, Ian Kirkpatrick and Grahame Thorne, with Martyn Sanderson as the All Blacks' coach and British actor Windsor Davies as the president of the Welsh Rugby Union.

I decided to start the film's international sales campaign by targeting Argentinean distributors. The Argentine rugby team had recently toured New Zealand and I thought this would give the film a profile in Latin America. I was wrong. Argentineans told me their rugby players never went to the movies, and Argentineans who went to the movies had no interest in rugby.

My most unexpected sale was to Japan, where the title was picturesquely translated as *No-Side They Scrummaged After 25 Years*. The Tokyo distributor prepared a subtitled theatrical release and sent me the elegant poster. I hung it on my office wall opposite the big French posters for *An Angel At My Table* and *The Navigator*.

As it happened, New Zealand and Japan were the only countries where *Old Scores* had a cinema release. I never saw reports on the Japanese box office, but in New Zealand 15 prints were released just before the Rugby World Cup. The timing was good but the results were less so: the film attracted only 15,000 New Zealanders, less than half the average attendance at a real test match. It was a different story when it was telecast on the British ITV network – 6.8 million people tuned in to watch.

We were starting to learn that theatrical markets were not always the best home for some of our films. But the evidence wasn't always accepted or even understood. Most producers and directors continued to hope for success in cinemas, even when it was suspected that some projects were 'not cinematic enough' because their scripts lacked sufficient revelation, tension, surprise, or the ability to reach their audiences' emotions.

There would be tougher lessons to come, as more features were completed and released. There would be bigger successes too.

TWELVE

MURDER, MELODRAMA
AND MIRAMAX

*N*EW ZEALAND WAS BACK in competition at the 1992 Cannes Film Festival with Alison Maclean's first feature. *Crush*, the story of a predatory American woman with whom an author and his young daughter become infatuated, had been filmed in the strangely exotic thermal region of Rotorua. Producer Bridget Ikin, hoping for future investment from Europe, had chosen London-based Carole Myer to handle international sales. I was torn between being pleased that a friend was selling the film and disappointed that, after the great success of *An Angel At My Table*, I wasn't doing it myself.

Carole hadn't wanted *Crush* to launch in the Cannes' competition, preferring the less pressured environment of the Directors Fortnight. She asked me to offer the first screening to Pierre-Henri Deleau: it was important the first contact came from New Zealand, rather than London, so the film's nationality wasn't confused. Deleau viewed the film and then – against our expectations – turned it down. Only then did I call Gilles Jacob, telling him the print was in Paris and available for him to see. By this time the competition deadline had passed, but insiders told us that although Jacob had screened more than 400 films he had selected only seven.

We were confident about getting into the competition because of the involvement of Pierre Rissient, an influential movie all-rounder for whom *eminence grise* would be a relevant description. He had been first assistant to Jean-Luc Godard on *Breathless*, a press attaché, a writer-director of two features, and then something more than a talent scout for international cinema. He had discovered Jane Campion's Australian short films, and looked after her when they were selected for Cannes in 1986. He was also an unofficial adviser to Jacob about films from the southern hemisphere.

Rissient had recommended *Crush* for the Cannes competition after visiting Auckland during the film's post-production. He was offended when he discovered we had given the print to Deleau first. Maclean – who looked on Rissient as a friend after his visit and his advice about editing the film – arrived at the festival to discover that communication between them had been cut off. We watched Rissient show his disapproval by crossing to the opposite side of the Croisette rather than come face to face with her or any other New Zealanders.

On the last night of the festival I accepted an invitation to the closing dinner, in the hope of re-establishing contact. Although I placed greater value on my direct relationship with Jacob, the decision-maker, I didn't want continuing problems with one of his advisers. Sighting Rissient in the crowd, I rushed across the restaurant and held out my hand so he had no chance of avoiding me. Communication was uneasily re-established.

Back home, *Crush* won four New Zealand film awards, including best actress for Caitlin Bossley, who played the author's daughter opposite Donogh Rees and American star Marcia Gay Harden, whose character was described in London's *Sunday Times* as 'a gorgeous witch … as sulphurous as the Rotorua springs'. The New Zealand release attracted 10,000 people, a smaller total than distributor John Maynard had expected. 'Our campaign focused on women, but the most influential critics are men,' he later rued.

Maynard's company was also the New Zealand distributor of Peter Jackson's third feature *Braindead,* which did three times the business. The horror-splatter-comedy didn't have expectations of selection for the competition, but after beginning its life in the Cannes market, with Jim Booth and Peter in attendance to oversee sales by Perfect Features' Grace Carley, it went on to win prizes in other festivals.

Braindead was advertised as the largest special-effects film ever undertaken in New Zealand or Australia. When it competed at Rome's Fanta Film Festival the jury – chaired by special-effects genius Carlo Rambaldi of *ET* fame – gave Weta Workshop head Richard Taylor his first international award for special effects. Tim Balme won best actor for his role as the increasingly demented Lionel, playing opposite Spanish actress Diana Penalver as his girlfriend. Penalver had been kept in the cast in spite of the loss of Spanish investment which had delayed the film's production.

Peter Jackson was being increasingly noticed. An Italian journalist dubbed him the infant prodigy of horror. At the Montreal Fantastic Film Festival, he collected best film and best director awards, and the audience waved sparklers and sang to celebrate his thirty-first birthday.

In the United States *Braindead* earned a bigger advance than *The Silent One*, setting a record for a New Zealand movie. Trimark's advertising campaign failed to convey the uniqueness of Jackson's style, but the lack of American theatrical success did nothing to slow his growing reputation as a director. Jackson, *Time* said, combined the legacies of George Romero and Buster Keaton. The film, though, was not universally praised. In *The New York Times*, Stephen Holden called it 'a promising concept run amok … proves that mad comic excess does not always ensure laughter.'

After winning the grand prize at the Avoriaz Film Festival, *Braindead* was a modest theatrical success in France, with more than 50,000 people seeing it in the first three weeks. *Le Monde*'s critic compared the film with a *Monty Python* show – at once hyperrealist and delirious.

Both it and *Crush* were selected for the Sundance Film Festival, more proof of the diversity of New Zealand production. Both also competed in the New Zealand film awards, with *Braindead* winning out with best film, best director and three other categories, including another best actor for Tim Balme.

Ironically, Peter Jackson was already planning his move away from the genre in which he had made his name. Continuing to live and work in New Zealand, he would complete his next feature film in less than two years. Alison Maclean, though, would join the diaspora, moving to New York where it would be seven years before she completed another movie – *Jesus' Son* (1999), based on Denis Johnson's short-story collection about a young substance abuser drifting through the fringes of American society. The cast, headed by Billy Crudup, included Samantha Morton and Jack Black, both of whom would later star in major New Zealand movies.

Leon Narbey's ambitious second feature *The Footstep Man* was also in the 1992 Cannes market. This densely complex film about the making of a film included scenes in a soundstage version of Toulouse-Lautrec's Paris, with New Zealand actors Michael Hurst playing the painter and Jennifer Ward-Lealand a prostitute. Producer John Maynard had asked Carole Myer to sell this film as well, but the persuasive Myer could not persuade a single theatrical distributor to buy it. It wasn't released anywhere except New Zealand, where attendances were under 3,000. With the aim of recouping some of her marketing costs, Myer transferred the film to a company which specialised in packaging features for sale to television.

As well as handling *The End Of The Golden Weather* and *Moonrise*, my sales responsibilities at the market included launching *Chunuk Bair,* a modest-budget first feature by Auckland-based brothers Grant and Dale Bradley. Based on a play of the same name by novelist Maurice Shadbolt, the film depicted New Zealand soldiers fighting against the odds to secure the highest point at Gallipoli – a horrendous military engagement in Turkey during World War I. The Bradleys, who believed in the value of stars from overseas, had hired English actor Robert Powell to play a New Zealand sergeant-major. They would continue to use the formula of importing offshore actors for lead roles. In the next 12 years their company, Daybreak Pacific, would produce almost 20 movies, often genre films starring American actors such as Kelly McGillis, Caspar Van Dien and Linda Parker.

Diversity continued to be the best way of describing New Zealand's film output. It certainly applied to a year when successful features included the dark and moody relationships of *Crush*, the blood-spattered comedy of *Braindead* and the warmly evocative nostalgia of *The End Of The Golden Weather.*

The successes were not, however, enough for the financially stretched National government to maintain the level of its support for film-making. In 1992 the Film Commission was coping with a 74 percent cut in its grant from taxation revenues – from $3.4 million to $888,888. Although lottery funding increased by $400,000 (to $7.8 million) this wasn't enough to stop the pain.

'As a sovereign nation, New Zealand cannot rely on others to tell our stories,' the commission's annual report remonstrated. A survey had shown that 88 percent of New Zealanders believed it was important for New Zealand films to be made.

Earnings from overseas sales of films in which it had invested became ever more critical, boosting the commission's diminished pool of money by more than $2 million. This, and co-production money from offshore, lessened the impact of the government cuts.

Our happy record at Cannes continued in 1993 when *Desperate Remedies*, a lavishly stylised nineteenth-century melodrama, was selected for the Certain Regard section of the official programme. The film, with the Auckland Philharmonia playing Verdi on the soundtrack, had been filmed in a wharf shed on the Auckland waterfront. It offered juicy roles to Cliff Curtis as a seducer, Kevin Smith as a lovelorn hero and Jennifer Ward-Lealand as a businesswoman with a secret.

Kevin and Jennifer came to Cannes to help with publicity. Kevin, a modest man with a spectacular physique, was often called on to take off his shirt before being photographed. Jennifer kept her character's ballgown hanging behind the office's kitchen door in readiness for photo calls. Between them, they carried out more than 20 interviews with international journalists.

The Auckland Philharmonia sounded wonderful through Cannes' immense sound system, but the levels – in response to our request to play the soundtrack loudly – were so high at the first official screening that some of the state-of-the-art speakers were damaged. We assumed repairs would have been necessary before the next day's screenings.

Before Cannes I had begun negotiations to license United States rights for *Desperate Remedies* to the high-flying Miramax company. The deal resulted from a Wellington screening attended by Tony Safford, who had changed companies since acquiring *An Angel At My Table* for New Line three years before. Tony's visit was supposed to be a secret, but it wasn't a well-kept one. The screening was also attended by cast and crew, and there was a buzz about the presence of an American buyer. Safford was unqualified in his enthusiasm, but the deal was conditional on his New York bosses viewing and approving the film. When I flew to New York the following week I took a print with me. The film got approval from Miramax's co-founder Harvey Weinstein and the deal was ratified.

A British deal was finalised just as quickly, and at the London Film Festival co-director Peter Wells introduced the first British screening in London's biggest cinema, the Odeon Leicester Square. The distributor was Liz Wrenn, whose Electric Pictures had grown to become one of London's leading specialty distributors. She had been disappointed three years earlier when I sold *An Angel At My Table* to Artificial Eye, one of her competitors, who had been able to pay a higher advance than she was willing to offer.

Her release was set for New Year's Eve, when the film would be an alternative choice to all the conventional Hollywood titles. It seemed a good strategy. But on New Year's Day there was a snowstorm and Londoners stayed home. The freeze lasted for three days. The release of *Desperate Remedies* never recovered. It closed after two weeks, even with reviews such as Stephen Dalton's in *New Music Express*: 'Gloriously overwrought, shamelessly camp and utterly wonderful ...'

Australian audiences – and Australia's weather – were more in tune. At the start of the film's four-month Australian release, Neil Jillett wrote in *The Age* that it looked and sounded like a collaboration between Charles Dickens, Baz Luhrmann, Henry Miller, Georgette Heyer, Graeme Murphy, Peter Greenaway, Barbara Cartland, the Goons and Jean Genet.

When the film had screened at Cannes, a red warning label had been stuck on every admission ticket informing the audience it contained scenes which might offend the sensibilities of some viewers. We assumed this referred to two sex scenes, and were bemused that French tolerance didn't extend to a film from New Zealand. The sex scenes also caused trouble with Japanese Customs officials when we delivered materials to our distributor in Tokyo. Pubic hair and sexual intercourse had to be cut, the border patrollers ruled.

The subject of censorship led to a drawn-out dialogue with Miramax, which seemed to be losing interest in its New Zealand acquisition. The company made it clear that cuts would be demanded by the American ratings system, and we waited for several months to be told what was required. Then, to the amazement of all, the film was given an R certificate without cuts. However, because of the film's lack of success in London, Miramax asked for some cuts to be made anyway. These would, it hoped, streamline the film for American audiences. Miramax also had an issue about one word. A reference to 'a town called Hope' had to be changed, it insisted, because Hope was the name of President Bill Clinton's home town and this might send the wrong message to American audiences.

After months of delays, Miramax gave the film a lacklustre theatrical release which began in the Quad, one of New York's smallest cinemas. *The New York Times* showed no mercy. Stephen Holden panned the film as 'a gorgeous but silly mock epic … aspires to be the '90s answer to a '40s Hollywood costume drama but with a big wink in its eye and layers of sexual ambiguity … Jennifer Ward-Lealand looks and acts like a cross between Greta Garbo in *Queen Christina* and Catherine Deneuve. Cliff Curtis's Fraser suggests the 1950s John Derek with too much make-up and a nipple ring.'

Desperate Remedies was released with more enthusiasm in Spain by Paco Hoyos, a film-lover whose influential distribution company had grown from his personal passion for movies. He had discovered the film four months before it was completed, when I showed him colour transparencies highlighting the lavish sets, costumes and actresses. We were in the New Zealand office at the American Film Market, and Paco was due to attend a reception at the Playboy Mansion. Instead he stayed in our office, fascinated by the images, until I persuaded him to sign a deal for a pre-sale.

Before its Spanish release as *Soluciones Desesperadas,* the film was selected for competition at the Sitges Film Festival. This festival was held in the new casino

Santa Barbara 1986.
The California city's first film festival included a 'Focus on New Zealand' attended by a delegation which included (from left) directors David Blyth and Peter Wells, Glen Rowling, actor Bruno Lawrence and the New Zealand ambassador, former prime minister Bill Rowling.

Cannes 1987. *Official signing of New Zealand's co-production treaty with France. David Gascoigne, Film Commission chair (seated right), watches as David Clement, director-general of the Central National du Cinema, signs. Standing from left: Christian Charret of the Centre National du Cinema, the author, and producer John Barnett, president of the New Zealand Independent Producers and Directors Guild.*

NZFC Collection, New Zealand Film Archive/Nga Kaitiaki O Nga Taonga Whitiahua

Cannes 1987. *A reception was held on the beach to honour the selection of* Ngati *for Critics Week. ABOVE: (from left) Carol Greene, Derek Malcolm, film critic of* The Guardian, *Danish distributor Frida Orhvik, and Sara Risher of New Line. BELOW: (from left)* Ngati's *director Barry Barclay, Yvonne Mackay of the Gibson Group, Jean Roy of Critics Week,* Ngati's *writer and associate producer Tama Poata, lead actor Wi Kuki Kaa, press attaché Søren Fischer, and Auckland lawyer Piers Davies who was in Cannes for a seminar on co-productions.*

New Zealand Film Commission

Cannes 1988. *On the balcony of the New Zealand office. From left: John Maynard, producer of* The Navigator, *which had been selected for the festival competition; Film Commission chief executive Jim Booth, soon to leave and become a producer; the author; and Barrie Everard, producer of* Never Say Die. *Fourteen years later, Everard would become the commission's fifth chair.*

Tokyo 1991. *Flowers and applause for* An Angel At My Table's *director Jane Campion (centre) and star Kerry Fox at a press conference before the Japanese release of the film, which had won eight prizes at the Venice Film Festival the previous year.*

Wellington 1992. *Director Ian Mune and Dr Diana Mason, widow of playwright Bruce Mason, arrive at Wellington's Embassy Theatre for the New Zealand premiere of* The End Of The Golden Weather, *based on a stage play by Mason.*

Cannes 1992. *Director Alison Maclean (right) became the second New Zealand director to have a feature film selected for the Cannes competition, with her first feature* Crush. *Arriving with her for the competition screening are American actress Marcia Gay Harden and British distributor Chris Auty.*

Sitges 1993. *In Spain for the official screening of* Desperate Remedies *at the Sitges Film Festival, Jennifer Ward-Lealand (centre), who won the festival's best actress award, and actor Kevin Smith (right).*

Cannes 1994. Once Were Warriors, *the first feature of director Lee Tamahori (centre), was a sensation at its market screenings during the 1994 festival. With Tamahori are four of the eight New Zealand directors whose short films were in official selection. From left: Grant Lahood* (Lemming Aid), *Jonathan Brough* (The Model), *Neil Pardington* (The Dig) *and Niki Caro* (Sure To Rise). *Nine years later Caro's second feature* Whale Rider *would set international box-office records for a New Zealand film.*

Wellington 1995. *The film industry invited politicians to hear why they should increase government support for New Zealand film-making. From left: Film Commission chair Phil Pryke, prime minister Jim Bolger and director Peter Jackson. Clips from Jackson's uncompleted fifth feature* The Frighteners *were shown.*

Cannes 1995. *Jeanne Moreau and Rena Owen meet at a beach restaurant during lunch at the festival. Italian and French reviews of* Once Were Warriors *had compared Owen's performance to that of a young Moreau.*

Wellington 1997. *The achievements of New Zealand film-makers were celebrated at a dinner held at national museum Te Papa Tongarewa to mark the 20th anniversary of the Film Commission. Since its founding the commission had financed more than 80 New Zealand feature films.*

LEFT: Director David Blyth (left) with producer Murray Newey. When their film Moonrise *was sold to the United States, its American distributors changed its name to* My Grandpa Is A Vampire.
BELOW: Rena Owen, star of 1994 box-office hit Once Were Warriors, *with Ramai Hayward, star of 1940 feature film* Rewi's Last Stand. *Ramai and her husband Rudall had travelled New Zealand screening and promoting his last film,* To Love A Maori.

Cannes 1997. *Danielle Cormack, star of* Topless Women Talk About Their Lives, *with Italian playwright and critic Guglielmo Biraghi. He had selected New Zealand films for two festivals he directed – Taormina and Venice. Both had won awards.*

Auckland 1999. *Film Commission chairman Alan Sorrell (left) and producer-director Gaylene Preston, first chair of the newly created Academy of Film and Television Arts, admire the silver-backed paua shell mounted on glass which would be presented to winners at the Academy's New Zealand film awards. The Academy, a pan-industry group, was set up to continue the awards after the collapse of the Guild of Film and Television Arts.*

hotel of an old seaside town half an hour south of Barcelona. As guests, Kevin Smith and Jennifer Ward-Lealand discovered Paco's predilection for all-night parties. Each night they were invited to join him for dancing and celebrating that went on until morning. Jennifer's reward came when the festival jury gave her the prize for best actress.

Although *Desperate Remedies* was the focus of our marketing drive at Cannes in 1993, New Zealand's biggest success of the festival would be *The Piano*. Jane Campion's film had been selected for official competition, and pre-sold for French distribution to a company run by Jean Labadie, who had scheduled his release to begin as soon as the festival was over. Driving from Nice airport to Cannes, I heard his radio commercials for the film on almost every station on the dial.

Produced by Australian Jan Chapman, *The Piano* had been fully financed by a French investor called Francis Bouygues. Bouygues had made a fortune in the construction business, building numerous notable structures, including the Musée d'Orsay in Paris, the European Parliament building in Strasbourg and Riyadh University in Saudi Arabia – at the time the world's largest building project. At the age of 68 he had embarked on a new career of feature film production, establishing a company he named Ciby 2000. Pierre Rissient was one of the advisers who helped him choose directors whose films he would finance. As well as Jane Campion, they included David Lynch and Pedro Almodovar.

The Cannes jury, chaired by Louis Malle, decided *The Piano* would be co-winner of the Palme d'Or with the China–Hong Kong co-production *Farewell My Concubine*. This made for two records: Jane became the first woman director to win the festival's top award, and Chen Kaige's film the first Chinese production.

Because *The Piano* was made by a Sydney production company, its official nationality was Australian. Nevertheless, everyone referred to it as a New Zealand film and I encouraged this, giving on-the-record comments at Cannes to all the journalists who wanted to write about it.

When the media interviewed Sue Murray, my opposite number at the Australian Film Commission, she disagreed with my patriotism, and headlines about the conflict appeared in the trade papers. It was a phony war. Both Sue and I were pleased to get the extra attention for the film. And after Cannes, Sydney's *Telegraph Mirror* saw the light, writing: '*The Piano* is as much a part of New Zealand as the kiwi or the haka.'

Jane Campion, who was pregnant, returned home before the festival ended. The organisers refused to allow producer Jan Chapman to go on stage to receive the award, telling her they wanted Sam Neill, not only because he was a New Zealander but also because of his popularity with festival audiences. Flown to France at the last minute, Sam pronounced the film 'a sort of miracle wrought by the superb Jane Campion'.

The New Zealand press, which seldom showed interest in New Zealand's successes at Cannes, had sent no reporters. When international news agencies reported that a New Zealand-made film had won the Palme d'Or, our office phones started to ring and we became unofficial correspondents for New Zealand newspapers and radio stations which hadn't bothered to do any advance planning. The media were better organised the following year when *The Piano* won Academy Awards: television coverage of child star Anna Paquin's victory speech was endlessly replayed.

Miramax's American release of *The Piano* grossed over US$34 million in its first five months, more than ten times as much as any earlier New Zealand feature had earned in the United States. By then the film had set box-office records in Germany and Italy, where Pandora and Mikado had each obtained the rights on the basis of their success with *An Angel At My Table*. Paco Hoyos was the Spanish distributor. Years later when I visited him in Madrid he was still playing the film's soundtrack on his car stereo system.

New Zealander Wendy Palmer, an accountant who had become one of the world's leading film marketers as head of a succession of international companies in London, was selling the film on behalf of its French financier. She decided Barrie Everard should release the film in New Zealand. He wasn't at Cannes, so she phoned him in Auckland and interrupted a dinner party he was hosting to tell him he had to buy it. When he said he would call her back in the morning, she said no. He had to decide there and then. So he said yes, although he hadn't seen the film. Wendy thus achieved her ambition of the film having a New Zealand distributor, instead of New Zealand rights going to an Australian company as often happened. The release was a big success.

New Zealand was even more visible at Cannes that year. We had two shorts in the competition: Glenn Standring's noir sci-fi *Lenny Minute*, and Grant Lahood's surreal black comedy *Singing Trophy*, which won a special jury mention for technical excellence. We hadn't been able to get seats for the closing ceremony, so we watched the live telecast of the presentation on television in our office. After some moments of disbelief, Grant Lahood's producer John Keir uncorked champagne to celebrate his win. We uncorked more champagne when Sam Neill accepted the award for *The Piano*.

A year later Grant was back in the Cannes competition with his next short film, *Lemming Aid. Sure To Rise* by new young director Niki Caro was also in the competition. The jury, headed by Clint Eastwood, awarded Grant the short-film jury prize. Jeanne Moreau presented it to him. More champagne was uncorked in the New Zealand office.

It was an exciting period for New Zealand short films and their keen young directors. 'I have an idea,' Gilles Jacob had written to me. 'Why not build a special programme with five or six shorts.' As a result, there were official Cannes screenings in 1994, of not just the two competition films but six others as well. My new colleague Kathleen Drumm repeated the short films in Los Angeles, inviting Hollywood producers and distributors whom she hoped could help the film-makers' future careers. Afterwards she toured the package in cinemas throughout New Zealand, with high-profile publicity benefiting from the Cannes selection. It was one of the rare occasions when short films found success in cinema release.

A cinema release proved more elusive for a new film we launched at the 1993 American Film Market. John Laing's *Absent Without Leave* starred Craig McLachlan, the blond, tanned, good-looking star of the top-rating Australian television soap operas *Neighbours* and *Home And Away*. The McLachlan factor helped me get a big video release in Australia, and a A$100,000 licence fee from Channel Nine. It also encouraged the BBC to buy the film.

At the time Craig was living in London, where he was to star in a West End revival of the popular '70s musical *Grease*. The BBC decided to use this to boost a cinema release of the film through the Rank Organisation. But after Rank market-tested the film it said the release wasn't worth risking. The British market, it said, was polarised; there was no longer any middle ground between studio blockbusters and art-house releases. *Absent Without Leave* fitted neither category, even with McLachlan's headline appeal to British tabloids. From that year onwards it became harder to sell New Zealand films to the United Kingdom.

The same polarisation was spreading throughout Europe. At the Mifed market at the end of the year, and at Los Angeles the next February, distributors interested in New Zealand films said they wanted only titles suitable for release in art-houses. Everyone else was looking for big-budget American movies made for the major studios. Hollywood was reasserting itself, extending its big films on to screens previously reserved for independent titles such as ours. Not only were outlets for independent films diminishing, but with the emergence of independent productions with American accents the market was becoming over-supplied.

This created problems for *Alex*, a sweet but modest story of a young New Zealand woman aspiring to swim at the Rome Olympics. A family film directed by Australian Megan Simpson and based on a book by award-winning children's author Tessa Duder, *Alex* was a co-production with Australia. When I sold it to one of the wealthiest Italian distribution companies, it earned a large advance. The company paid in full, and took delivery of materials for making the Italian language soundtrack. But we never heard from the company again and it never released the film.

There were always exceptions of course. I was able to sell a revenge thriller with supernatural elements to more than 50 countries. *Jack Be Nimble* was a first feature for director Garth Maxwell and his two producers, former *New Zealand Herald* journalist Jonathan Dowling and distributor Kelly Rogers. The plot revolved around a brother and sister seeking revenge against parents who had abandoned them as babies. American actor Alexis Arquette made a convincing New Zealand school-boy, complete with shorts, long socks and kiwi accent. A third-generation member of an American acting family, Arquette was dressed in readiness for a drag cabaret act when Garth took me to meet him in Los Angeles. We drove him, bewigged and begowned, from his house to the cabaret, where he gave a brief and witty performance.

David Overbey of the Toronto Film Festival called *Jack Be Nimble* the best film debut in years. When the film opened briefly in London, critic Alexander Walker wrote in *The Evening Standard* of 'a vividness that recalls David Lynch in *Blue Velvet* mode'.

Stephen Holden was equally complimentary in *The New York Times*: 'This Gothic horror film has hallucinatory power and psychological refinement that you won't find in any number of Stephen King novels … What makes it a superior genre film is that it never loses its focus on childhood trauma and obsession and their tragic repercussions.' But the British and American distributors seemed to focus little attention on film's release, and in spite of its good reviews it wasn't widely seen.

Although Miramax had, similarly, proved unenthusiastic about its release of *Desperate Remedies*, it would show unqualified confidence in another New Zealand film – even before it was completed. *Heavenly Creatures*, Peter Jackson's fourth feature, marked a change for the director from splatter-horror comedies to a story of wider audience appeal – although the fantasy elements of his early film-making would still surface on the screen.

The film was based on a true story – the murder by two teenage girls of one of their mothers in the staid southern city of Christchurch in the 1950s. Co-scripted by Jackson and his partner Fran Walsh, it was co-financed by the Film Commission and Senator Film Produktion, a company owned by Hanno Huth, a German producer and distributor who had released Jackson's *Braindead*. Huth, who put up two-thirds of the money, persuaded the commission to give him the rights to sell the film everywhere in the world except New Zealand, Australia, Hong Kong and Taiwan, territories the commission would retain.

Early in 1993 when the film was in production, Huth phoned and told me he was finalising a substantial deal to license all his 'foreign' territories to Miramax. He recommended I do the same.

I discovered Miramax executives had been given a script and were coming to New Zealand to view the work in progress. It seemed the American distributor was so keen on the film that negotiations with the notoriously aggressive company would not be difficult. Like all its competitors, Miramax had been following Peter's career since *Bad Taste*. Could he create a film which would appeal to larger numbers of people than just horror fans? This, the company had decided, would be the one.

When Tony Safford phoned me to say Miramax wanted to acquire all rights for the four territories I was selling, we began negotiating at once. The negotiations gathered momentum at the 1993 Cannes Film Festival and it didn't take long to agree on the final advance. However the wording of some clauses in the draft agreement provided by Miramax's lawyers proved more difficult to resolve. Final negotiations kept me in Cannes after the festival had ended as we haggled about definitions and percentages.

While this was going on, I was also finalising the deal for Miramax to buy North American rights for *Desperate Remedies*. I had promised co-directors Stewart Main and Peter Wells and their producer James Wallace that they would meet Harvey Weinstein. But Weinstein was always too busy to meet us. And I wasn't seeking to meet him in connection with the simultaneous *Heavenly Creatures* negotiations. In fact, after hearing stories – often involving a lot of shouting – from other sellers I wasn't anxious to confront him at all.

One night when I was asleep in my Cannes apartment, my bedside phone rang and a silky voice was on the line. It was about 2 a.m. Harvey Weinstein was calling. He had decided to use his considerable powers of charm and persuasion to resolve an issue on which the negotiations for *Heavenly Creatures* had been foundering. I had heard about his habit of using home phone numbers. One of his staff had told me how Weinstein insisted he install a special phone line in his home apartment. The line was never used except by Weinstein, so he could always get through at any time, from anywhere.

I marvelled at the smoothly cajoling voice. Harvey Weinstein was legendary for his stormy negotiating but this was a different technique. Now, more than ten years later, I cannot remember what the issue was. I only remember the promises of how well Miramax would handle Peter's film. Four days before we reached final agreement on *Desperate Remedies*, Tony Safford and I signed a deal for Miramax to acquire the commission's four *Heavenly Creatures* territories. With the New Zealand deal and the deal with Hanno Huth, Miramax now controlled all of the world for the uncompleted feature, except for the United States and Canada which Huth hadn't yet sold, and Germany and Austria, which he was keeping for himself.

Two months after Cannes, Tony Safford came back to New Zealand with his New York colleague David Linde. Behind the usual Miramax screen of confidentiality, I assumed they had obtained the right to see *Heavenly Creatures* before it was shown to any other American company. I also assumed the film would be premiered at the next Cannes festival. Indeed Pierre Rissient had said a slot was being kept for it.

In April the following year Miramax announced it had acquired the rights for the United States and Canada. By that time – again to ensure the film's New Zealand nationality was made clear – I had been asked to oversee last-minute arrangements to send the first print to Paris, and was faxing Gilles Jacob to tell him when he could screen it. However Harvey Weinstein soon intervened and started his own dialogue with Jacob, one which ended without agreement.

Jacob offered a midnight premiere at Cannes, but out of competition – an offer Miramax rejected. Instead the film was held back for five months, finally screening in competition at the Venice Film Festival. Miramax had added its logo to the front of the film and was able to claim a share of the glory when the jury, headed by David

Lynch and including Nagisa Oshima and Uma Thurman, awarded the film a Silver
Lion. The prize, New Zealand's second major one at Venice, brought the second
standing ovation for a New Zealand director. *Variety*'s critic presciently wrote that
Heavenly Creatures was 'exhilarating and original. Jackson's drop-dead command
of the medium positions him to be catapulted far beyond his sci-fi and splatter
excursions.'

At the Toronto Film Festival the film won the Metro Media Award. The *Globe
and Mail* called it truly mesmerising. When it opened in the United States, *Film
Comment* proclaimed it a masterpiece, and *Time* named it one of the year's ten best
films. British reviewer Christopher Tookey wrote in *The Daily Mail*: 'Astonishingly,
under the influence perhaps of his co-writer Frances Walsh, Jackson has turned out
a film unrecognisable from the rest of his oeuvre – mature, sensitive and profound.'

The year after Venice, a jury headed by directors John Carpenter and Tobe
Hooper gave it the best film award at the Paris Fantastica Film Festival, the same
(re-named) event where *Bad Taste*, with a fifteenth of the budget, had won the special
jury award seven years before. And in Los Angeles, Peter Jackson and Fran Walsh
were nominated for an Academy Award for best original screenplay.

Heavenly Creatures begins with a dedication: 'For Jim'. Jim Booth, the producer,
had been diagnosed with cancer, and although he had kept working as long as he
could, he had died early in 1994 before the film was completed. *Heavenly Creatures*
was his third production for Peter Jackson since leaving the commission five years
earlier.

In Venice, Peter said the award was 'truly a tribute to Jim Booth'. For the Academy
Awards, he had prepared an acceptance speech in which he would pay tribute to the
man whose five-year plan had included 'Win an Academy Award'. 'We didn't win,'
Jackson said later, 'but it was close enough.'

Miramax, which would license *Heavenly Creatures* to more than 40 countries,
asked that the film be shortened for its international releases. The company sent
suggestions, but promised that the final decisions could be made by Jackson and
Walsh. Ten minutes were subsequently removed under their supervision.

Delivery of the final international version was overseen by a cheerful young New
Yorker named Charlie McClellan, who was Miramax's associate director of post-
production. Miramax sent him to New Zealand to meet the director, and I took him
to dinner at the Brooklyn Café and Grill, then one of Wellington's best restaurants,
for his first meeting with Jackson. They got on very well. Within a year, Charlie had

been hired by Peter and had moved to Wellington as digital effects producer for his next feature.

At the same time as it was acquiring *Heavenly Creatures*, Miramax announced a first-look deal for future projects written and directed by Peter Jackson and his production company WingNut Films, and for projects originated by Fran Walsh. The deal, though, didn't result in any Miramax productions of Peter Jackson films. Instead Jackson made his next feature, *The Frighteners*, for Universal Pictures, with Robert Zemeckis – director of all three *Back To The Future* movies and *Who Framed Roger Rabbit* – as executive producer and Michael J. Fox as lead actor.

'It's a little bit of Hollywood coming to New Zealand; it's not me going to Hollywood – that's something I've always resisted,' Jackson told *Onfilm* magazine.

The Frighteners became New Zealand's biggest-earning summer release, with a box office of $1.5 million; it also earned another special-effects award for Weta Workshop's Richard Taylor at the Sitges Film Festival. However in the United States it did less than the blockbuster business the studio had expected.

Universal had announced that Jackson's next feature for them would be a remake of the legendary cult film *King Kong* – a longtime dream of the director. But following the release of *The Frighteners* the studio seemed to cool on the idea and the project ground to a halt. Instead, Jackson returned to the Miramax fold and started to develop *The Lord Of The Rings*, after the company obtained the film rights from veteran American producer Saul Zaentz.

As development of the *Lord Of The Rings* project continued, Miramax and its new owner Disney grew tense about the rising budgets and Jackson's wish to make a trilogy. This culminated in a last-chance stand-off: Jackson negotiated a three-week period when he could search for a new production company. At New Line in Los Angeles he and Fran Walsh met Bob Shaye and his colleagues, who were prepared to finance three features. Fifteen months of shooting began in October 1999.

Although the hugely successful trilogy was not a Miramax production, Harvey Weinstein and his brother Bob succeeded in getting executive-producer credits, plus a percentage share of the gross box office, because the project had been initially developed with their money. Like the Weinsteins, Jackson would also benefit considerably from a share of the earnings. He would plough millions of dollars of this into his Wellington infrastructure.

With gross box-office earnings in the billions of dollars, all three *Lord Of The Rings* films are today in the list of the world's top ten international earners. The trilogy won a total of 17 Academy Awards, including best film and best director for the final film, *The Return Of The King*. In New Zealand the first film, *The Fellowship Of The Ring*, holds top position in the box-office charts, with ticket sales totalling $14.5 million, followed by *The Return of the King* with $13.2 million.

Only four feature films have exceeded $12 million at the New Zealand box office. One is *Titanic*. The other three were made by Peter Jackson.

«Un film coup de poing
qui vous laisse émus
et bouleversés»

L'ÂME
DES GUERRIERS
UN FILM DE LEE TAMAHORI

THIRTEEN

BIGGER THAN
JURASSIC PARK

*L*EE TAMAHORI'S FIRST FEATURE *Once Were Warriors* starred Temuera Morrison and Rena Owen in the best roles of their lives. But it nearly didn't get made at all.

When producer Robin Scholes brought the project to the commission, we were all excited that Lee had at last found a feature he wanted to direct. He had been in the film industry for more than 15 years, starting as a boom operator on eight features (the first was Geoff Steven's *Skin Deep*) and then becoming first assistant director on eight more productions, including Oshima's *Merry Christmas, Mr Lawrence* and Geoff Murphy's *Utu* and *The Quiet Earth*.

An enthusiastic and knowledgeable film-goer, he was an admirer of mainstream action directors such as Sam Peckinpah and Sergio Leone. But he had ended his feature-film involvement in 1986 and moved on to lucrative success as a maker of television commercials – directing more than a hundred and winning awards in the United States and Australia. During this time he had directed a half-hour film titled *Thunderbox* for producer Larry Parr, but had not found a feature he wanted to make, until *Once Were Warriors* offered him a story with 'a lot of elements about New Zealand that were of interest to me – social politics, high drama and very interesting characters'.

In spite of publicity which said the film was about a subject nobody wanted aired, commission staff were pleased the best-selling first novel by writer Alan Duff would be filmed. The book had sold more than 30,000 copies since its publication in 1990. But it had aroused heated controversy because of its intense story of violence in a Maori family.

Scholes, who had been working in television after returning home from film studies at Edinburgh University, believed the film would be controversial in a positive sense because violence in society was real and needed to be faced. 'Until we put these images on screen,' she said, 'people will not have the chance of openly discussing the issues.'

Alan Duff was asked to write the screenplay. After he delivered an early draft, the commission's development committee refused further financial assistance. Judith McCann and head of development Ruth Jeffery encouraged consideration of another writer, arguing that a woman's perspective would help make the film more acceptable for cinema audiences. The film-makers' choice was Riwia Brown. Brown was experienced as an actor, writer and director in Maori theatre but had never before tackled a feature film. Working with the producer and director, she restructured the script, strengthening the role of the wife, Beth, and making her the principal character in place of her violent husband Jake. Tamahori would later say that Brown had added 'a lot more hope, heart and positive things, without destroying the violent core.'

However there was another hurdle to be overcome. Although staff believed the commission should invest in the production, board members were not persuaded. They were nervous about whether a film on a violent domestic relationship would appeal to theatrical audiences. They turned down the first application for production money.

Two months later Robin and Lee brought the project back to the board. This time they brought a group of Maori leaders with them. Each spoke about why it was important for the film to be made. Lee also spoke persuasively about his vision. It would be a tough, gritty drama that would address serious problems we had as a nation. And the movie's characters would reflect the urban lifestyle of contemporary kids and speak their language. This time the board said yes.

To create the feel he wanted, Lee chose people who, like him, had been working in the industry since the end of the '70s. Director of photography Stuart Dryburgh had been the cinematographer on *Vigil* and *An Angel At My Table* and been nominated for an Academy Award for *The Piano*. Designer Mike Kane had worked with Lee on *Utu* and *The Quiet Earth*, and had designed many of his commercials. Kane and Dryburgh collaborated to give the film the dark, saturated colour the director wanted. Dryburgh referred to this as an enhanced reality.

Also enhanced was the physique of the leading man. Temuera Morrison, an experienced film and television actor who had starred in his first movie as a child in 1973, spent three months increasing his bodyweight and bulk. His training programme took him from 74 kilograms to 83.

The film was perfectly timed for the 1994 Cannes Film Festival. In preparation, I sent a print to Paris, where Michelle Seawell added French subtitles to ensure the colloquial dialogue would be understood by the festival's selection committee.

The committee, however, failed to invite it for any official section of the festival. This wasn't what we had expected, but there was no time to argue or negotiate. Our response was to set up our own premiere in the Cannes market. Admission was by invitation only, and our invitations made it clear this was a special event.

When I arrived in Cannes, I discovered that British sales agent Carole Myer had scheduled a feature at the same time and was targeting many of the same distributors. Invitations had been sent out, so it was too late for either of us to change. The screenings took place in the same multiplex and we both had full houses. However some of the key buyers, including the Australians, chose the British film. I sat outside our cinema on the steps of the big foyer, talking with a New Zealand television crew. Carole stayed inside her screening, her presence near the door ensuring no one would dare to leave. I watched the audience come out of the British film. It was obvious they had not been impressed.

Our audience came out a minute later. They were in a highly emotional state. Many were weeping. Others were hugging each other. All were talking about the intensely moving experience. Lee, Rena, Temuera, Riwia and Robin had introduced the film, and were waiting as the audience came out. As I watched the distributors embracing Lee and Rena I knew I had never seen an audience so affected by a film.

By the second screening the following day, word had spread. Distributors were turned away. But most of the Australians saw the film – they arrived early to be sure of getting seats. While the final credits were rolling, Lyn McCarthy and Graeme Tubbenhauer, co-owners of the Australian Dendy distribution company, left the cinema and ran the several hundred metres along the Croisette to our office, where I was talking to Alan Finney, the only Australian who didn't seem to have heard of *Once Were Warriors* and its potential. When Lyn and Graeme arrived panting at our door, they stared in dismay at their competitor, thinking he had been negotiating to buy the film. In fact Alan departed without mentioning it. The Dendy representatives came in. Lyn asked what advance I wanted. I told her. This was another rare occasion when there was no negotiation. They agreed to pay the price.

They had made an astute decision. The film grossed over A\$6 million in their 60-print Australian theatrical release and millions more in the video release, earning more than Quentin Tarantino's *Pulp Fiction*.

In New Zealand, where it was contracted for release by John Maynard's distribution company Footprint Films after some of the bigger companies had turned it down, it was equally successful. By this time John was based in Sydney, but his New Zealand company had maintained its commitment to New Zealand movies.

Everyone agreed that such a controversial production should be handled with caution, so the film began its New Zealand release – at the same time as we were launching it at Cannes – with only five prints. The first week's results were huge and the local critics were universally enthusiastic. *Pavement* editor Bernard McDonald called it 'enthralling and impressively moving … a film which moved me to tears.' Seventeen more prints were ordered for the second week. There were no more doubts about how audiences would respond to the film.

Once Were Warriors became a national obsession, even more than *Goodbye Pork Pie* had been 15 years earlier. It was in New Zealand cinemas for 12 months and was seen by more than a million people – six times as many as *Heavenly Creatures*, and more than either of the previous local record-holders *Footrot Flats* (730,000) and *Goodbye Pork Pie* (603,000). It out-grossed *Jurassic Park* and *The Lion King* to become the first film from anywhere to sell more than $6 million worth of tickets at the New Zealand box office. In the New Zealand film awards it won eight prizes, including best film, best actor, and best screenplay for Riwia Brown.

The New Zealand television premiere on TV3 in prime time attracted 700,000 people – 52 percent of all viewers. The channel showed it without cuts – four-letter words and all. It was, said programmer Bettina Hollings, 'an icon of New Zealand culture'.

An interesting story lay behind the private broadcaster's acquisition of television rights. Avalon Studios, a subsidiary of public broadcasting company Television New Zealand, had invested in the film before production began on the recommendation of its head of investment Sue Thompson, who'd been given the script by Larry Parr, then Avalon's head of production. The deal included television rights, but when Avalon offered the licence to its parent company, TVNZ executives turned it down, citing among other things the amount of swearing in the script.

The rejection caused problems: before finance could be obtained from television funding agency New Zealand On Air – a necessary part of the budget – the producer had to prove a broadcaster was signed up to screen the film. And Avalon Studios needed income from a television sale to start recouping its investment.

Fortunately, film director Geoff Steven had become a programming executive at TV3. Steven proved to be more perceptive than his public counterparts. On his recommendation, the commercially owned channel agreed to acquire the television rights. And, knowing the film had nowhere else to go, Steven negotiated a fee which would prove to be far below the market value of such a successful film.

One day after the first Cannes screening of *Once Were Warriors*, Jim Murphy of Malofilm persuaded me to license the film for Canada. I resisted his offer at first. I hadn't yet discussed the film with any of the United States distributors at the festival, and had always believed that deals for the larger country should be finalised before negotiating with its less populated neighbour. My thinking had no doubt been influenced by occasional threats from American distributors saying, 'If we can't have Canadian rights we won't buy US rights.'

However this time I took a punt and signed a deal with Jim before any Americans turned up. The experience taught me that there were advantages in separating rights for the two North American nations, not the least being that income was earned from two sources instead of one.

There had in fact been earlier interest from the United States, but I was keeping quiet about it. After finalising the two deals for Miramax the previous year, Tony Safford had come back to New Zealand and become the first offshore buyer to see *Once Were Warriors*. It impressed him so much that we negotiated a deal for Miramax to acquire it. However, the deal was subject to approval from Harvey Weinstein.

Commission staff kept the secret as a print was sent to New York. The decision was phoned to me when I was on my way to France: Weinstein thought the film was too violent for American audiences. He wouldn't go ahead. In these circumstances, another condition came into play: no one was to know we'd done a favour for Miramax by showing it the film before any other company. Three Miramax staff turned up to our Cannes premiere, notebooks in hand, pretending it was the first time their company had seen the film.

New Line's Mark Ordesky was the next American to identify the film's potential. Four years earlier Tony Safford had been the negotiator when New Line acquired North American rights for *An Angel At My Table*. Now Ordesky acquired its second New Zealand feature, but only for the United States.

Once Were Warriors had its North American premiere at the Montreal Film Festival. Dennis Hopper was head of the jury that awarded it best actress – worthy recognition for Rena Owen, whose performance had not been rewarded at home – and best film. It also won the Ecumenical Jury Award for giving new hope to victims of violence. From then on, whenever I met people who were uneasy about the film's violence, I told them about this prize – awarded by three clergymen.

Rena Owen and Lee Tamahori were travelling from Montreal to the Telluride Film Festival in Colorado when they received the news that prizes were to be awarded to them; they had to retrace their journey – three flights – to get back for the presentation. From Montreal they flew to the Venice Film Festival, arriving two hours before the official screening, where the film earned another standing ovation for New Zealand and the award for best first feature.

The festival was then being run by Gillo Pontecorvo, whose masterful feature *The Battle Of Algiers* had featured in my first Wellington Film Festival in 1972. The great Italian director was now 75. After we had invited him to see *Once Were Warriors* at Cannes five months earlier, he had become a fervent admirer of both the film and Rena Owen's performance.

American reviews of *Once Were Warriors* were exceptional. *Time*'s Richard Corliss called the film 'another stunner from the bustling New Zealand film community' and chose it and *Heavenly Creatures* as two of the ten best films of the year. *New York* magazine's David Denby said, 'Let's hear it for New Zealand. This obscure, under-populated, once-pacific country has made a film that couldn't possibly intersect more crucially with our bitterest preoccupations.' *The New York Times*' Janet Maslin described Temuera Morrison and Rena Owen as 'frighteningly credible because they both look like pure sinew and brawn'.

New Line's Fine Line division kept the film in theatrical release for more than four months, and its box office was around US$3 million – about the same as for *Heavenly Creatures*, but less than the company had expected. Key staff changed at the time, and the film's advertising seemed undecided about what audiences were being targeted. Both things may have contributed to the disappointing result. Another reason could have been word of mouth: many Americans told me they thought the film was too violent. I always challenged this, arguing that Hollywood films had for decades delivered much worse violence. Oh yes, the Americans would say, but your violence is *real*. Ours is only fantasy.

After *Once Were Warriors* had begun its United States release, I picked up the phone in my Wellington office one day to be told that French film star Gerard Depardieu was on the line from Los Angeles. He had seen the film and wanted to praise 'the actors, the story and the emotion'. He asked if the film were available for release by his French distribution company. But he was too late. I had already licensed rights for France, where 200,000 people would see the film in its first four months.

Once Were Warriors was becoming the biggest cinematic success New Zealand had ever experienced. In 18 months I had negotiated contracts for the film to be released in 60 countries – the largest number of sales that I'd made for one film. The final total would reach 100.

Hanno Huth acquired all rights for Germany and released 55 prints for a theatrical release lasting five months. The film ran for nine months in Italy, more than six in Scandinavia, seven in South Africa and four in London, and became one of the few New Zealand features released in Singapore, where a reviewer wrote in entertainment magazine *8 Days* that 'it leaves you crying and your knees weak'.

Before *Once Were Warriors*, Rena Owen had been unknown in the film world. She had worked for nine years in theatre and television in Wellington and London, and had written a play. Her only experience in a feature film had been a small part in Kevin Costner's production of *Rapa Nui*, shot on Easter Island, where most of her scenes had ended on the cutting-room floor. Now her role as Beth Heke would bring her international fame.

For the first releases, Rena travelled with Lee Tamahori to the United States, the United Kingdom and Australia, her visits creating immense media coverage for the film. She later toured Italy, Germany, France, Spain, South Africa and Japan. In France, *Le Monde* critic Henri Behar called her the Antipodean equivalent of Anna Magnani. In *Vogue*, John Powers wrote: 'Tamahori draws an electric performance from Rena Owen, a farouche beauty who stretches her huge mouth into great randy smiles. Almost ecstatic in her victimisation, she gives Beth's suffering a histrionic grandeur that many audiences find emotionally overwhelming.' Britain's *Empire* magazine said she had delivered a performance of towering proportions.

A year later I invited her back to Cannes to meet many of the international distributors with whom she had worked, and who were having so much success with *Once Were Warriors*. At lunch one day at a beach restaurant, Jeanne Moreau was at a nearby table. *The New York Times* had said Rena had the sad, sensual look of Moreau, so I encouraged her to introduce herself to the woman who had been at the top of French cinema and theatre for more than 40 years. As it happened, Moreau had seen the film. It was, she told us, an honour to be compared with Rena.

Given the acclaim for Rena Owen's performance in *Once Were Warriors*, everyone expected her to win the 1994 New Zealand best actress award. It therefore came as a shock to us all, but especially to Rena, when the judges instead awarded the prize to Australian actress Genevieve Picot for her lead role in Gaylene Preston's mini-series *Bread and Roses*, which had been released in New Zealand cinemas in the short time between its completion and its television broadcast.

In North America, this stirring story of the early life of gutsy New Zealand trade unionist Sonja Davies premiered at the Vancouver Film Festival, where a local critic called it a compelling, beautifully realised tribute to a generation of New Zealand women. In Melbourne, *The Age* said Picot gave probably the finest screen performance ever by an Australian woman. But apart from film festivals – including Seattle, where it was voted the festival's 'sleeper hit' – its theatrical life didn't extend outside Australia and New Zealand. I had thought the series might follow the same path as *An Angel At My Table*, but this did not happen. Apart from two 16-millimetre prints, money was never found to create theatrical materials. International sales were handled by an Australian television company, which licensed it for telecasting in more than 30 countries.

The following year Gaylene Preston extended her focus on 1940s' women with her feature-length documentary *War Stories*. This moving and unusual work consisted of seven elderly women, one of them Preston's mother Tui, talking with exceptional bravery and candour about their lives and love affairs during World War II. The women were filmed by Alun Bollinger against a plain black studio backdrop, with a single interviewer who was never seen. Their personal stories were intercut with newsreel footage. The release by Lorraine Steele for John Maynard's distribution company made *War Stories* the top-grossing New Zealand feature release of its year, running for more than five months with a total audience of 60,000. There were cheers at the Sydney Film Festival when it was voted number one documentary. It also earned selection at Sundance, Venice and Toronto, and beat Peter Jackson's *Heavenly Creatures* for the award of best New Zealand film of the year.

Preston's achievement in persuading her subjects to talk so openly about intimate aspects of their lives was followed by a bonus when she flew six of them (the seventh having died) to Los Angeles for the United States premiere at the American Film Institute, followed the next day by afternoon tea at the house of comedian Phyllis Diller.

Gaylene often argued that a full-length documentary was just as much a feature as a dramatic fiction film. True as this is, documentaries have seldom attracted large cinema audiences. I negotiated an American cinema release for *War Stories*, but despite admiring reviews the film didn't resonate with American audiences and was seen by few people.

Four years later it was the same when I found a United States distributor for Annie Goldson's feature documentary *Punitive Damage*, a moving story about a New Zealand mother investigating the murder of her son in East Timor. The film won awards in Munich and Sydney, and was widely admired when Annie and I attended its premiere in New York. But the theatrical release was a disappointment.

The film didn't find an American audience until it was telecast by Home Box Office.

American cinema audiences would, indeed, prove elusive for many of New Zealand's 1990s feature films. When John Reid made *The Last Tattoo* in 1994, he designed it to be attractive and accessible to American audiences. The ambitious romantic thriller, set in Wellington (but actually filmed in Dunedin) during World War II, starred Tony Goldwyn, grandson of the legendary independent Hollywood producer Samuel Goldwyn, and veteran American actor Rod Steiger, as well as New Zealander Kerry Fox who had earned an international profile in *An Angel At My Table*. The film's American sales agent pre-sold it throughout most of Europe, but in spite of the American stars and the efforts of co-producer Bill Gavin, who made three trips to the United States in search of a distributor, it joined the list of movies that would not get an American cinema release.

The same fate befell the first feature directed by Tony Hiles. The film came with promising credentials: Peter Jackson – with whom Hiles had worked on *Bad Taste* – was executive producer, co-writer, special effects supervisor and second unit director. Because of the Jackson connection, Hanno Huth had agreed to invest. His new Berlin-based Senator Film sales agency, helped by a brochure which misleadingly placed Peter's name above the title, pre-sold the film to 16 countries, and he released it in Germany. But neither the Jackson factor nor the quirky premise – an inventor with a medieval monk trapped inside his head – was enough to generate interest from American distributors.

The lack of interest continued at home, where fewer than 3,000 people went to the film. Despite the small audience, *Jack Brown, Genius* had its admirers in the industry: at the New Zealand film awards it won best director, best actor for Tim Balme as the inventor, and best score.

Producer John Keir and I succeeded in persuading Huth to invest in one more New Zealand film. We met him at Cannes and showed him Grant Lahood's two award-winning short films, which made him laugh. As a result, he agreed to invest in the director's first full-length feature, a black comedy called *Chicken* – with the commission, as co-investor, obligingly giving him worldwide sales rights. But when the film was completed Huth and his sales staff didn't like it. At a mutually embarrassing meeting, they told us they would not handle the film. Then they returned the sales rights to the commission. I succeeded in licensing it to only two territories: Korea and Poland.

This was the end of Huth's four years of investment in New Zealand features, and later he closed his sales agency.

Gregor Nicholas's second feature *Broken English* was a more successful title from this period, and one of the few which got United States theatrical distribution. Sony Pictures Classics released his story of a young Croatian woman trying to forge a new life in New Zealand. It was welcomed by Janet Maslin in *The New York Times*: 'An impassioned film … ably conveys the full range of confusion created in New Zealand's large and varied influx of immigrants.' Its British release brought praise from *Sight and Sound*: 'Despite a star-crossed-lovers storyline that's anything but fresh, it shows that sheer charisma can make shop-worn material absorbing.' At home it was seen by more than 100,000 people and won five awards.

Another New Zealand feature – albeit with a dual nationality – which had a release in the United States was directed by Jane Campion's sister, Anna. *Loaded* (originally called *Bloody Weekend)* was made as a British-New Zealand co-production, with John Maynard and Bridget Ikin as the New Zealand producers and the commission contributing a third of the budget. Shot in England, where Anna lived, and post-produced in Wellington, the film was sold to Miramax, who briefly released a United States version 12 minutes shorter than the original prints. The British Film Institute, which had sales rights, shared the opinion of *Variety* that 'commercial chances look grim'. The predictions proved correct.

This was the first time the New Zealand Film Commission had invested in a film in which neither New Zealand nor New Zealanders were visible onscreen. A British critic, however, managed to find a rationale. In the *Financial Times*, Nigel Andrews opined: 'No two cultures are more compatible than those of New Zealand and the United Kingdom. Same language, same repressive wholesomeness, same ability to find nightmares under the tea table.'

Co-productions would always prove challenging. Such films have to try and maintain a clear national identity, despite involvement and influence from other countries. While financially beneficial for producers – in future years they would make bigger-budget productions possible – co-ventures have to skirt the danger of being seen as films from nowhere. *Loaded* didn't have this problem: it was clearly seen and heard as a British movie – not a relevant result for its New Zealand investors.

Two New Zealand-Canada co-productions would have identity problems of a different kind. Wayne Tourell's *Bonjour Timothy* and Ian Mune's *The Whole Of The Moon,* both targeting youth audiences, were shot in Auckland and post-produced in Montreal, where co-producer Micheline Charest had found investment from Canadian television funds. A successful and ambitious television producer, she aimed to move into feature films for the cinema.

She was also a strong negotiator, and another who persuaded the Film Commission to give her control of all sales rights. This worked to the disadvantage of the films when her Canadian-based sales agents insisted on launching them at television markets. Both directors had made their films for theatrical release, and indeed the commission's policies were to invest only in films for cinemas. But by the time I had persuaded the Canadians they should screen the films for cinema distributors, it was too late: both had become branded as telemovies. Micheline later told me there was another reason they garnered little theatrical interest: it was, she said, because of television influence during post-production at her company's television facilities in Montreal.

My part in the films' international promotion was, frustratingly, limited to ensuring that both got a chance to win international awards – always one of our strategies to make New Zealand films stand out from the crowd. *Bonjour Timothy* won awards at Berlin, Giffoni and Belfast, and *The Whole Of The Moon* at Berlin and Giffoni. Despite this, the films had cinema releases only in New Zealand, where *Bonjour Timothy* – a light tale of an Auckland schoolboy falling in love with a Canadian exchange student – was the more popular. *The Whole Of The Moon*, however, won more New Zealand film awards – five, including best film.

Murray Newey, the New Zealand producer of both films, intended to make a third co-production with Micheline Charest, to be shot in Canada and post-produced in New Zealand. But it was never made. The commission was reluctant to invest in another film in which New Zealand wouldn't be seen, and which was chiefly aimed at television audiences.

The two co-productions were the last features Newey would produce, although by then, in partnership with John Barnett, he had acquired the rights to Witi Ihimaera's story *Whale Rider*. The two producers started working on the film with Ian Mune, who wrote the first script and was expected to direct it. However almost ten years would pass before *Whale Rider* was made – for a different company and with a different director.

Par le réalisateur du Seigneur des Anneaux

Peter Jackson
nous a joué
un tour digne
d'Orson Welles
L'ÉVÉNEMENT DU JEUDI

Forgotten Silver

un film de Peter Jackson

un film écrit et réalisé par Peter Jackson et Costa Botes · Avec Leonard Maltin, Sam Neill, Harvey Weinstein, Lindsay Shelton, Marguerite Hurst, Peter Jackson, Costa Botes, Michael Shortland, John O'Shea, Johnny Morris

cheyenne films

FILM OFFICE

FOURTEEN

TAMPERING WITH
SUCCESS

*O*NCE WERE WARRIORS AND *HEAVENLY CREATURES* became such high-profile
international success stories, with the clamour continuing for two years
beyond their first screenings, that the features which followed them in the
mid 1990s struggled to earn similar attention. In the event, two completely different
films came to share the limelight. Each ran for less than an hour. Each was received
with a level of enthusiasm usually restricted to full-length productions. And each
earned the elusive prize of cinema release in the United States.

The most talked-about was *Forgotten Silver*, co-directed by Peter Jackson with
fellow Wellingtonian and former critic Costa Botes. The film was ostensibly about
the amazing discovery of the lost works of a pioneer New Zealand film-maker called
Colin McKenzie. It included fabulous extracts from the film-maker's movies, with
Jackson talking earnestly about their unearthing. McKenzie, said Jackson, had filmed
the first flight, made the first colour-film stock, and built a set for a legendary city for
a film with 15,000 extras.

In fact, the whole thing was a clever spoof. The fabulous extracts from the imagin-
ary pioneer's films were all directed by Jackson himself. The story was concocted by
him, Botes and Fran Walsh, and filmed just before he became totally involved with
The Frighteners. Harvey Weinstein, Sam Neill and American critic Leonard Maltin
all made guest appearances to give credibility to the tall tale. The 'world premiere'
of one of the lost masterpieces had been filmed before a session of the Wellington
Film Festival. Acting the part of the festival director, I had made a speech introducing
the film (which was, of course, not shown) while Botes cued the festival audience
to applaud.

The day after *Forgotten Silver* premiered at Wellington's Paramount Cinema, it
was broadcast on national television. Cooperative journalists had written previews
as though the story were true. When the prank was revealed, viewers were furious.
But Botes was unapologetic. 'If *Forgotten Silver* causes people never to take anything

from the media at face value, so much the better,' he said in response to a flood of angry letters in the newspapers. 'The sense of humour kept shining through,' media commentator Brian Priestley wrote loftily in *The Christchurch Star*. 'Can it really be true that thousands of people took the thing seriously?'

The commission was so pleased with its investment in the film that it held a special screening in the cramped theatre of Parliament's executive building, the Beehive, with Peter Jackson on hand to answer questions. The Jackson name attracted plenty of members of parliament. Later it would attract plenty of buyers for the film from all over the world.

Forgotten Silver was selected by the Venice Film Festival, who screened it one day after *The Frighteners*. It was popular with North American festivals too. When it launched in New York, Lawrence Van Der Gelder wrote in *The New York Times* that it managed 'simultaneously to position its hero in the path of great events, while sending up its subject, film history, with informed skill, great affection and mischievous glee.'

The second 50-minute film to draw worldwide attention was *Cinema Of Unease*. Actor Sam Neill had been chosen by the British Film Institute to make a documentary reviewing the history of New Zealand film – part of an international series commemorating the centenary of cinema.

Neill's international career after *Sleeping Dogs* had included *Jurassic Park, The Hunt For Red October* and more than 30 other movies and television mini-series. He had worked all over the world, but retained strong links with New Zealand and made frequent unpublicised visits to his house in Wellington and his vineyard in the South Island. He engaged another expatriate, Sydney-based documentary-maker Judy Rymer, to write and direct the film with him.

Cinema Of Unease, which included clips from dozens of New Zealand movies, was the only film in the series to be directed by an actor. Jean-Luc Godard directed the French film, Stephen Frears the British, Nagisa Oshima the Japanese and Martin Scorsese the American. These four films, plus Neill's, were selected for the Cannes Film Festival and then released in the United States by Miramax.

Neill cleared several months in his acting diary to come home and direct the film, which he told a Los Angeles news conference was 'prestigious, but a complete folly to do'. He also appeared in front of the camera as narrator, saying that in the virtually comatose '50s suburbia of his childhood, the pain of isolation was salved by the Saturday matinees. All this, he felt, had been changed by the production of New Zealand movies. 'What intrigues people about New Zealand cinema is that it's so dark. It has that brooding quality. There's always the potential for violence.' The title reflected this opinion.

I found it interesting to compare local views of the film with those of overseas critics. Wellington Film Festival director Bill Gosden found the film a trip back in time because of its 'once familiar case history of Anglophiliac cultural cringe'. But he wrote that 'the vivid scenes from New Zealand movies – good, bad and wonderful – remind us how empowering it has been for New Zealand audiences to see New Zealanders on the big screen.' When the documentary screened four months later at the New York Film Festival, *New York Times* critic Janet Maslin saw it as underscoring 'the strain of madness and savage rebellion that has poured out of New Zealand ever since that country stopped making upbeat travelogues and started developing a strong film industry.'

While New Zealand's film industry was being seen overseas as vital and energetic, at home a new Film Commission chair was deciding some dramatic changes were necessary. Australian-born investment banker Phil Pryke had made his reputation advising the New Zealand government on sales of public assets such as Telecom, the Coal Corporation and Postbank. He was appointed the commission's third chairman on April 1, 1993, after having been a board member for less than a year. He took over from David Gascoigne, who had held the position for eight years and been a board member since the commission began.

Pryke signalled an immediate change of style when he announced that, except for the executive director, the commission's staff would no longer attend board meetings. This ended the close working relationship between senior staff and board members that had, in my experience, been a strength of decision-making. The productive and constructive relationship had been exemplified in the outgoing deputy chairman, advertising executive Bob Harvey, who had an ability to listen to new ideas from the staff and then support them with authoritative enthusiasm. Pryke's board included people who had worked closely with senior staff in many capacities, including *Variety* correspondent Mike Nicolaidi and Wellington producer Dave Gibson. But with the new separatism, dialogue and debate between board and staff – even between old friends – seemed to be no longer possible.

The new chair announced that he was reviewing the industry and overhauling the commission; he began meeting film-makers to canvass their views. The commission's sales activities were included in his overhaul. It seemed he wanted to close down sales altogether. The subject was discussed at meetings of producers attended by Pryke and the commission's executive director Judith McCann. Secrets are hard to keep in a small industry: film-makers frequently phoned to give me their version of what was happening and to ask why, as the person responsible for sales, I hadn't been at the meetings. I asked Judith if I could attend, but the answer she relayed from the new chair was no.

I believed that the commission's sales record was a strong one and that there were definable benefits from selling the films which we were also financing and marketing. I met with the chair and stressed my belief that it would be bad for the industry if the commission stopped selling. I told him how our sales campaigns promoted the whole industry, and how our promotional work was another way of generating sales. I gave him a list of 12 producers who had chosen the commission to sell their independently financed films. In response, Pryke advised me there would be a second round of meetings. This time he agreed I could attend.

There appeared to be no consensus about stopping sales. While some producers, including John Barnett, Bill Gavin and my former colleague Jim Booth, spoke in favour of finding sales agents overseas, others including Robin Scholes, Murray Newey, Trevor Haysom, Gaylene Preston and Larry Parr supported the existing system. Parr said he liked the fact that the commission's earnings went back into the industry to be reinvested in more films.

The meetings were followed by a draft discussion paper written by two board members – Dave Gibson, who had closed his sales company five years before and placed his films in the hands of the commission, and Pauline Hughes, an Auckland public relations executive. Their draft included options for discussion, but these didn't include the existing system of sales by the commission. When I told the chair I was disturbed by the omission, he said he hadn't realised the status quo had been left out and would ensure it was added. But it wasn't included when the final version of the paper went out to the industry. Producers again sent their responses. Again there was no substantial agreement on change.

After Pryke had been in the position less than six months, Judith McCann was stunned to read a press interview in which he announced that her job was to be re-structured, re-named and re-advertised. Although he had told her he was considering these changes, he hadn't said he intended to announce them. Soon afterwards she had a phone call from Australia, asking if she would be interested in becoming chief executive officer of the South Australian Film Corporation. After an interview in

Adelaide, she decided to accept the offer. New Zealand's loss was Australia's gain. Her five years at the commission had included investment in three of the most successful films it had ever supported. McCann would come to be admired for rejuvenating South Australia's film industry.

Her successor in the rewritten position of chief executive was Richard Stewart, an Australian who had been head of Film Queensland in Brisbane. The appointment of an Australian to run a New Zealand cultural body raised some eyebrows, and did nothing to calm an atmosphere of growing suspicion and argument. In comparison with McCann, who had worked late nights and weekends writing board papers and providing detailed analysis for each investment proposal, Stewart proved to be more hands-off. With his arrival, the burden of paperwork moved to other desks.

Soon after his arrival there was a board meeting which generated so much noise staff members came out of their offices to discuss what might be happening. From outside the closed door of the boardroom we could hear angry shouting; it was a startling change from the collaborative attitudes encouraged during the previous 15 years. We learnt later that the new chief executive's strategic plan and first budget had been thrown out. Staff increases, restructuring and a new office in Auckland had failed to get support.

It was in this climate that the board agreed in mid 1994, at a time when *Once Were Warriors* was becoming the most successful film we had ever financed or sold, that the commission should stop selling New Zealand movies. To me, the decision was incomprehensible. Pryke and Stewart had been at Cannes that year, and had seen how the commission's office had been crowded with international distributors because of the success of *Once Were Warriors*. They also knew the commission had earned a six-figure sum from my sale of the four *Heavenly Creatures* territories.

Nevertheless, Stewart told me the issue was beyond discussion. He might, however, have been having some doubts. The industry journal *Onfilm* quoted him acknowledging that the industry was split down the middle. 'I don't want to be blamed as the person who closed down something which was a recognised success,' he said. He added, 'If the new system doesn't work, we could always bring back the old one.'

I persuaded the chair to allow me to speak to the board before the final decision was taken. I gave four brief reasons for opposing the proposal. Then I left the room, and after a brief debate the chair got what he wanted. The board decided that over the next three years the commission's sales of feature films would be phased out.

Stewart asked me to write a press release. It wasn't difficult to do – I was over-familiar with the anti-sales arguments – but I felt a certain irony in being asked to draft the official announcement of a decision that I had so strongly opposed, and that would result in a large part of my job being abolished. My draft was approved by the chair without comment and released on July 11, 1994. In it, Pryke announced that the commission was moving out of the business of selling films. Producers, he said, would have to find overseas sales agents able to commit finance for a film before the commission would consider investing. He believed this would somehow encourage producers to secure more finance from a wider range of sources and thereby improve the quality of their films.

The proposal bore a remarkable similarity to the policies of five years earlier. Then, in the face of similar funding requirements, production had languished. The same thing was about to happen again.

Word of the proposed changes soon spread. The business columnist of the British Film Institute's monthly magazine *Sight and Sound* wrote in September: 'Latest and most depressing convert to the supposed brave new world of commercial cinema is New Zealand, which recently cancelled the sales remit of its Film Commission, telling producers they should either sell their own films or, better still, enter into pre-sales arrangements with (probably US) sales companies.'

The commission's marketing department was, it said, one of the most cost-efficient operations in the spendthrift business of film, building up a sales record for small titles that no European country could equal.

'In Europe, where state bodies that sell as well as promote films are the exception, every small film institute and commission has repeatedly argued that the money spent on promotion could be more effectively used in helping to sell the films as well.'

Dave Gibson, who had been on the board since Pryke became chair, accused me of being cynical and bitter about the decision. 'I'll always be cynical,' I told him. 'But it's not correct to call me bitter.' Bemused was more like it. The policies had proved their cost-effectiveness in earning revenue while also promoting New Zealand and its film industry. In comparison, I had watched countries such as Canada, France and Germany spending huge amounts publicising their films without having any films to sell.

Ripples spread quickly through the industry. Former board members John Maynard and Bridget Ikin, who had produced such movie successes as *Vigil*, *The Navigator*, *An Angel At My Table* and *Crush*, said the new policies lost sight of the fact that the business of film in New Zealand was grass-roots rather than corporate, and that its survival rested on cultural not economic imperatives. In protest, they closed their New Zealand production companies. Four years later Maynard would also close his distribution company, which had released 14 New Zealand movies, an unequalled record.

The effect of the no-sales policy could have been far reaching. For 15 years, whenever New Zealand films had been officially selected for large international festivals, the commission had represented the industry at the screenings, not only negotiating sales but also generating interest in New Zealand's film-making potential. We had proved that sales and promotion fitted together. But such were the passions of the time that permission to travel for this purpose was now refused.

I asked the management committee for approval to go to Montreal, Toronto and Venice for the official festival screenings of *Once Were Warriors* and *Heavenly Creatures*, but the proposal was rejected on a split vote. When the films won their awards, New Zealand had no official representation. By the closing night at Venice, Peter Jackson had gone on to another festival in another country, and investor-producer Hanno Huth had also left. It fell to Sue Rogers, Jim Booth's partner, to collect the award for *Heavenly Creatures* – an emotional moment for her so soon after Jim had died.

Fortunately, many sales had been completed before the festivals began. Associated Press reported from Toronto that 'New Zealand may be the envy of everybody here. Both its films have United States distribution deals and the films are the most talked-about works at the festival.' At Venice, the Australian Film Commission took pity on the visiting New Zealanders and invited them to an Australian reception.

Although the chair wanted the commission to do less at overseas film markets, the period would, ironically, include a decision to spend more money at Cannes. Pryke had been uncomfortable with the modest size of our office during his first visit. The budget was increased and we moved into a larger space. And though we were supposed to be stopping selling, we were not stopped from attending markets. So the annual routine continued, taking me next to Mifed, where another retrospective, *Il Cinema Nella Lunga Nuvola Bianca*, supported by the city of Milan, showed 30

New Zealand features and shorts, and featured the first speech in Italian from the newly appointed New Zealand ambassador, Judith Trotter.

Pryke's first term as chair ended after three years. In his last annual report he acknowledged the successes of *Once Were Warriors* and *Heavenly Creatures*, which he said had generated more worldwide attention for New Zealand than the America's Cup and the Rugby World Cup. He also referred to the closure of the sales agency, which he claimed was 'in the spirit of co-venturing to ensure our dollars go further'.

In fact the closure hadn't happened. Plans were never made for any hand-over of the 50 features in the feature-film catalogue. There had been off-the-record discussions with a few overseas sales companies and other advisers, but nothing resulted. With each film owned by a different production company, and each with an individual sales agreement, the task of changing sales arrangements would have been complex, prolonged and costly.

By the end of the wind-down period, the subject seemed to have been forgotten. The sales catalogue continued to grow. The commission continued its promotional campaigns positioning New Zealand as the source of innovative film-making talent. It also retained its customary flexibility: producers who wanted or needed offshore sales agents were free to appoint them.

Jim Booth had used this flexibility to have Peter Jackson's second and third features sold by Perfect Features. But his choice had disappointing consequences. During the last year of Jim's life, when he was losing strength because of his spreading cancer, his concern grew about unauthorised expenditure by the British company, and delays in reports and payments. He was disheartened that the agency didn't seem to be meeting all its obligations. After Jim's death, Peter terminated the deal. Following some legal skirmishing we got access to the sales records from London, and *Meet the Feebles* and *Braindead* joined *Bad Taste* in the commission's sales catalogue.

Pryke was replaced as chair of the commission board by Auckland barrister Alan Sorrell. A board member for three years and deputy chair for one, Sorrell brought a calming voice to the boardroom, even if some film-makers found his style less sympathetic than they had hoped. Within a few months of Sorrell's appointment, in an apparent and welcome reversal of the earlier decision, the board was studying a paper about re-energising sales.

Richard Stewart resigned after three years of his four-year contract, and returned home to Queensland. Alan Sorrell promised industry input into selecting his successor, whom he said was most likely to be a New Zealander. When the choice was Ruth Harley 'everybody cheered her appointment', David Gapes declared in *Onfilm*.

Harley had been the founding executive director of New Zealand On Air – an independent body set up by the government to finance New Zealand production for both public and private broadcasters. Starting work at the commission, she discovered 'a grumpy industry' and promised to work to raise its low morale. She defended the commission's sales role, saying it was uniquely suited to New Zealand's circumstances, and noted that the commission's sales brought higher returns than offshore sellers', with their more expensive cost structures. She pointed out that when there had been a period of increased sales by offshore sales agents, returns to New Zealand had been much lower.

But the debate would never end. In 2003 the report of a Screen Industry Task Force, with John Barnett as an influential member, would revive the call for the commission to stop selling films.

FIFTEEN

SERIAL KILLERS, DEMONS, VAMPIRES AND THE UN-DEAD

*T*WO MORE HORROR FILMS were big earners during my last five years selling New Zealand movies. Both were first features, both were written by their directors, and both directors had made their reputations with short films. Their success re-fuelled the debate about which New Zealand films were most likely to succeed internationally – and whether investors such as the Film Commission should be selecting projects by trying to assess market and audience demand, rather than choosing from the stories film-makers wanted to tell.

Scott Reynolds' *The Ugly* began its sales life at my 17th Cannes market in 1996, before it was completed. Working with the film's producer, Jonathan Dowling, I finalised a Japanese pre-sale to Shochiku, a distinguished studio with a history going back to silent cinema. Shochiku's contact with New Zealand had begun when it released first *Braindead* and then *Heavenly Creatures*. After reading the script and viewing a five-minute videotape of scenes from the film, company representatives decided they wanted to release *The Ugly* as well.

The 28-year-old director had learnt about films while working as a teenage projectionist at the Hollywood Cinema in suburban Auckland, where one of his earliest memories was watching *The Incredible Shrinking Man*. He had stood out from the crowd of short-film directors with his 16-minute *Game With No Rules*, also produced by Dowling. Starring Danielle Cormack, Marton Csokas and Jennifer Ward-Lealand, it had been one of the six New Zealand short films chosen for the 1994 special programme at Cannes.

Reynolds' feature was ready for its first screenings at the American Film Market in Los Angeles early in 1997. Stylishly brutal and scary, it delivered 90 minutes of speedy tension in which a blonde psychiatrist, played by Rebecca Hobbs, tries to decide if a serial killer has been cured, but instead finds he is destroying her sense of reality. The killer was played by Naples-born Paolo Rotondo, who had come to New Zealand at the age of eleven and had later trained in acting, writing and directing with Paris-based Philippe Gaulier. Jennifer Ward-Lealand played his evil mother.

The pre-sale to Japan was a good omen: in the end I sold the film to 40 more countries, including a United States deal with Trimark, another of *Braindead*'s distributors. Trimark initially felt the film wasn't strong enough for Americans, but after we let them know that their competitor Miramax had told us the film was 'too scary and too dark', they made their first offer. As with *Braindead*, American crowds didn't flock to the cinemas to see it, but the video release sold a large number of copies.

Six months after Cannes, I took a detour to Spain on my way to the Mifed market in Milan and joined Scott Reynolds at the Sitges Film Festival, where his film had been selected for competition. He was a cheerful breakfast companion at the festival's casino hotel, and became even more cheerful when the jury awarded him the prize for best director. A month earlier, at the Fantastic Film Festival in Rome, *The Ugly* had won more prizes: best actor for Rotondo (whose fluent Italian appealed to the organisers of the festival, who invited him to be a guest), and best screenplay for Reynolds. Later in Portugal it won best actress for Rebecca Hobbs.

The film received admiring reviews wherever it showed. The influential French magazine *Mad Movies* (which would five years later publish a 132-page issue about Peter Jackson) wrote: 'We only get one fantasy film at this level in any year, so *The Ugly* is a rare commodity. It is from New Zealand, a country that regularly distinguishes itself with talented film-makers in this genre.' London's *Time Out* magazine concurred: 'It's a pleasure to see a film that uses bold formal ideas rather than expensive special effects to achieve its ends.'

Three years later a movie with the unlikely title *The Irrefutable Truth About Demons* became the next New Zealand horror film to make an international splash. It starred 27-year-old Karl Urban in his first lead role, playing a sceptic who is attacked by a weird satanic cult whose existence he refuses to accept – until he's painfully forced to acknowledge there are ways of looking at reality which he has never accepted before. Director Glenn Standring – whose short film *Lenny Minute* had been in the Cannes competition in 1993 – had graduated from the University of Otago with first class honours in archaeology, before moving to Canterbury University's Ilam School of Fine Arts (where Vincent Ward and Gaylene Preston had been students) to complete a film degree. He had made his feature during six weeks of night shoots in Wellington – using, he said, every dark alley in the capital city.

Distributors in France and Benelux signed up to buy the rights for their territories before the film was completed, again on the basis of script, publicity and promo reel. The film was even more popular when we started to screen it in the Cannes market in 2000. I went on to sell *The Irrefutable Truth About Demons* to more than 50 countries. The North American premiere took place at the Toronto Film Festival, where programmer Colin Geddes acknowledged 'the bloodline of New Zealand horror films pulsates in Standring's dark cautionary tale'.

We received three offers for United States cinema distribution. Two were from theatrical distributors. The third was from Blockbuster, a wealthy video company, which offered three times as much money if we would give up our theatrical ambitions and allow it to give the film its first release on video and DVD. The issue of forsaking an American theatrical release was delicate. The director and his producer, Dave Gibson, spent much time discussing it with me. Finally we decided we would take the money. The earnings would help the film recoup more than its budget in less than a year. And we would still have the right to screen the film in North American festivals, where critics would continue to admire it.

Blockbuster had made the offer without seeing the completed film: this was another deal that was driven by the script, the brochure and the promo reel, which featured computer-generated special effects from Gibson's First Sun production company. When the film was finished, Dave took a print to Blockbuster's Dallas headquarters, where he screened it in their private cinema. The film lived up to the company's expectations and we signed the deal.

In the week the film was released I was driving to a meeting in Los Angeles when I saw a huge Blockbuster sign above a store a few blocks from the beach in Santa Monica. I drove into the supermarket-size car park and walked into the store. Hundreds of tapes and DVDs of *The Truth About Demons* – the Americans had removed 'irrefutable' from the title – were massed on the shelves in front of me. I imagined the numbers multiplied by every one of the chain's 8,000 stores, and decided the film's American audience would be easily as large as if it had been released first in cinemas.

In spite of all the international sales, *The Irrefutable Truth About Demons* was seen in New Zealand cinemas by only 15,000 people. This was three times as many as had seen *The Ugly,* but both results showed that home audiences continued to be slow to discover that their own film-makers could make scary films just as well as Hollywood. New Zealand film-goers – or at least the ones who went to horror films – retained a strong predilection for horror with an American accent.

As with *Death Warmed Up* and *Bad Taste*, the Film Commission had been the sole investor in *The Ugly* and *The Irrefutable Truth About Demons*. People often asked me why a public body was willing to finance horror films. The answer was simple: it had every chance of both making a profit and helping launch new talent. Peter Jackson's career was, of course, the most convincing example. Commission investment in Peter's first five films – two of them horror comedies, the other three with substantial fantasy elements – had totalled around $5 million, and enabled him to prove his potential as one of the world's most successful film-makers. His next five features, no longer needing commission backing, brought production spending of several hundred times this amount into the New Zealand economy.

This was just the start of the benefits for the country. Jackson's share of *The Lord Of The Rings* trilogy's enormous international box office enabled him and his colleagues to create a film-making infrastructure in Wellington which they would make available to the whole industry.

From the time David Blyth completed *Death Warmed Up*, international distributors had keenly accepted New Zealand horror films as more than able to hold their own. It is surprising, therefore, that to this day so few have been made. Blyth and Jackson made one each in the 1980s. In the 1990s, after *Braindead*, only Reynolds and Standring followed Jackson's lead. And like him they continued their careers in New Zealand. Reynolds made two thrillers with producer Sue Rogers – *Heaven*, completed a year after *The Ugly*, giving Karl Urban his first feature-film role, and *When Strangers Appear*. Both were produced with offshore finance.

Standring sold his Wellington house and moved south to Dunedin to prepare for a fantasy vampire feature with producer Tim Sanders. Titled *Perfect Creature*, it was completed in 2005 as a co-production with the United Kingdom.

In the first years of the twenty-first century, only one other horror film was made, when 29-year-old Greg Page converted part of a rural landscape into 'a heartland of horror' for *The Locals*. Less gorily explicit than its predecessors, the story of the un-dead on a Waikato farm nonetheless earned sales to more than 40 countries. It was another title that bypassed cinemas and was released first on DVD and video in North America. Its United States distributor, Anchor Bay, had already discovered New Zealand talent: Peter Jackson's *Bad Taste* was in its catalogue.

New Zealand's other neglected genre is science fiction. Geoff Murphy's *The Quiet Earth* was one of its year's top foreign earners in the United States and sold to more than 80 countries, yet it remains the only one of its kind. Distributors often told me that New Zealand was a logical setting for science-fiction movies, and audiences and critics seemed to agree, but the genre has not been pursued by any other local film-makers. Perhaps they have been intimidated by the gigantic budgets of the Hollywood versions.

There were other examples of the industry failing to learn from experience. In the mid 1990s, when she toured the world to promote *Once Were Warriors*, Rena Owen became the most famous face of New Zealand cinema. She was photographed in Italian *Vogue* and interviewed in leading North American magazines. *The New York Times* likened her to Bette Davis and Jeanne Moreau. The mass-circulation *Movieline* magazine said she delivered an electrifying performance. *The Toronto Sun* praised her 'awesome screen presence'.

In an industry that needed its own stars with international recognition, Rena was the most internationally acclaimed New Zealand cinema actress of the decade. Despite this, she gained a leading role in only one other New Zealand feature, playing a 40-something singer returning home to reassess her life in Garth Maxwell's under-appreciated 1998 drama *When Love Comes*. When the film premiered at the Toronto Film Festival, I watched Canadians stopping Rena in the street to tell her they remembered her performance as Beth four years before.

Otherwise she was seen only in two small roles in New Zealand features. In Athina Tsoulis's *I'll Make You Happy* she played a lesbian prostitute, and in *What Becomes Of The Broken Hearted?*, the sequel to *Once Were Warriors*, she reprised her role of Beth, but only in a few brief scenes. Other feature roles were discussed. None eventuated.

The extraordinarily talented Bruno Lawrence had a similar experience. After playing iconic roles in *Smash Palace*, *Utu* and *The Quiet Earth,* and lesser characters in ten other 1980s movies, he seemed set to become an international name enhancing the market value of New Zealand productions. Yet in the years before he died in 1995, apart from a role in Garth Maxwell's *Jack Be Nimble*, he acted largely in Australia. After his death Lawrence was honoured with a Rudall Hayward Award for his outstanding contribution to a film industry which hadn't been big enough to find ways of continuing to use his talent.

Alongside *The Ugly* in the 1997 Cannes market, we featured the strangely named, laconically amusing *Topless Women Talk About Their Lives*. The film's director Harry Sinclair had studied drama at École Philippe Gaulier, had been one half of the legendary Front Lawn multimedia performance group, and had created several memorable shorts, including the circular and fascinating *Lounge Bar,* which he co-directed and co-wrote with singer Don McGlashan. In it Sinclair played a lovelorn barman who temporarily loses his memory when hit on the head by a falling bottle, and regains it when hit by a bottle for a second time.

Sinclair shot *Topless Women* in an unconventional way: over 25 weekends. Seeking spontaneity from his actors, he wrote and delivered each segment of the script only a day in advance. Thus when star Danielle Cormack became pregnant in real life, he was able to include her pregnancy in the storyline.

Sinclair and his producer Fiona Copland – both making their first feature film – persuaded the Film Commission to invest in *Topless Women* without a complete script, something it had never been willing to do before. They did this by showing a series of four-minute *Topless Women* episodes that Sinclair had made for TV3. These brief stories of contemporary life in New Zealand's biggest city featured the cast – including Joel Tobeck, Ian Hughes and Willa O'Neill – who would also be in the movie. All had been chosen by Sinclair as similar to the young Aucklanders they were to portray.

The episodes offered a perfect way of marketing the film before it was completed. After screening several at the Mifed market, including an outrageously comedic story involving sex in a toilet, I finalised pre-sales for the film to be released in Germany and Italy. But, like Peter Jackson with *Braindead,* the film-makers weren't happy. They argued that buyers should see the completed film first. I tried to get them to change their minds, even persuading Carole Myer to phone them from Milan to explain the many advantages of seeking deals before films were completed, on the rare occasions when this was possible. But Harry and Fiona were resolute.

Their stance would reduce the earnings of their film. I was close to pre-selling all rights for the United Kingdom when their ban took effect. The potential British distributor changed her mind when she saw the completed movie. It wasn't what she had expected, she said, and she decided not to buy it after all. The Italian and German pre-sales went ahead, however, and although the Italian release was small, the film had a 30-print cinema release in Germany. It also sold to Canada and Australia, and to a Japanese distributor who became a fan of Harry Sinclair movies and signed up to release his second and third films as well.

At home, the film was a popular success. After premiering in the Auckland and Wellington Film Festivals, it won best film, best director, best actor and best actress at the film awards and was seen by more than 64,000 people. But it never found United States distribution. Too raunchy, perhaps.

It was a different story with Sinclair's second film, *The Price Of Milk*. We launched this at Cannes in the same year as *The Irrefutable Truth About Demons*, headlining the fact that Karl Urban – who had just completed three leading roles with the Auckland Theatre Company – was starring in both. In the charming and imaginative *The Price Of Milk* he played the rural lover of Danielle Cormack on a fantastical CinemaScope dairy farm.

New York distributor Jeff Lipsky hadn't been on the list of people we contacted to promote our screenings. He had just started a new company named Lot 47 Films, and we hadn't yet heard about it. But he had seen our advertising and our brochures – they were piled high in all the main hotels and cinemas – and as a result he was in the audience at the film's first market screening. The following morning he walked into our Cannes office, shook my hand and said he wanted to buy *The Price Of Milk* for release in the United States.

Harry was sitting on the balcony, looking out at the crowds on the Croisette and no doubt worrying whether his film would be in demand. I asked him to come inside for a meeting, where Jeff told him how much he loved the film. Danielle and Karl were also at Cannes, and when they came back to the office they too met the distributor, who immediately started to talk about his release plans.

Jeff was so enthusiastic he flew to Auckland two months later to join us at the New Zealand premiere in the Civic Theatre, where he said, 'New Zealanders seem to lack understanding of just how potent, how brilliant and how powerful their films are.' Then he and his staff began to create an unflagging American publicity campaign, which included a sweepstake with prizes of trips to New Zealand.

It seemed logical to call Air New Zealand to give them a chance to get involved. We needed four return tickets to New Zealand as prizes; in return we offered exposure in all the film's American advertising on television and radio and in newspapers and cinemas. But the airline turned us down, making it curtly clear they couldn't see any point in being involved with the release of this New Zealand movie.

It was very different when I phoned Qantas. The Australian airline was quick to give us tickets, and to accept all the free advertising we were offering. The agreeability of their Los Angeles staff even extended to removing the kangaroo from the logo which they gave us to use in the film's advertising, in deference to the fact that this was a New Zealand and not an Australian movie.

Jeff Lipsky later wrote to me: 'I have seldom seen such in-house camaraderie and such a unanimous feeling of pride and joy in working on a film as I am seeing among my staff on *The Price Of Milk*. No one ever tires of seeing the film, of talking about, selling it, or becoming as angry, frustrated and vengeful as I am about the fact that so many critics have failed to grasp its wonderment and splendour.'

The New York Times at first wrote favourably. 'The film,' it said, 'uses the sparseness of New Zealand's great rural expanse to cast a classical fairy tale in modern terms … the eerie beauty of the countryside allows the plot laden with magic realism to work.' This was in an article by Andrew Johnston a week before the release, when the paper included the film in its recommended list. When, however, it was reviewed on Valentine's Day, which Jeff had chosen as the first day of its release, Stephen Holden dismissed it cruelly as 'terminally whimsical … a contemporary fairytale that goes increasingly haywire'. Lipsky boosted his campaign with postcards, cowbells and pint-sized milk containers filled with *Price Of Milk* candy hearts, but attendances fell short of his expectations.

It was the same in New Zealand, where the release attracted only half the numbers who had seen Sinclair's first film. *The Price Of Milk*'s overseas festival life was more encouraging: it was praised as enchanting at Toronto and Edinburgh, won best film awards at fantasy film festivals in Tokyo and Puchon, and special jury awards at St Tropez and Oporto. At home, its vibrant cinematography won recognition for Leon Narbey at the New Zealand film awards.

Like Jeff, I was puzzled and disappointed when the film didn't deliver the results we all expected. In the programme for the world premiere, Bill Gosden had written that the film reasserted the glories of the giant screen – 'movie stars, true love and betrayal, a magnificent landscape' – and had acknowledged the 'spinning loveliness' of Cormack and the 'smouldering charisma' of Urban. All of this should have been enough to draw crowds. Perhaps the elements of magic realism and unconventional fairytale were too much for everyday film-goers. Or did the film's bright surface not disguise Sinclair's darker vision of New Zealand as a 'make-believe country that seems almost empty of people, where love stories are played out in silence and loneliness … where life seems perfect but we find ways of making ourselves unhappy'?

Some people had a more pragmatic view, arguing that the narrative structure would have been stronger had Sinclair written a complete script before he began shooting, instead of continuing his *Topless Women* style of writing and shooting episodically.

On the surface, though, *The Price Of Milk* exuded sexual joyfulness and confidence, making it an exception in an industry where – as Sam Neill had observed – many films had shown a bleak, pessimistic sensibility. Nine months after her appointment, the commission's new chief executive Ruth Harley had suggested – or perhaps hoped – the industry had the energy to develop a feel-good stream alongside its art-house stream. She told journalist Phil Wakefield in an *Onfilm* interview: 'People like laughing. So let's do a bit more of that.' Her critics complained she wanted nothing but breezy comedies, a charge she denied. Rather, she said, she was all for diversity and quality.

In search of more cheerful topics, Harley resuscitated a co-financing deal that had been developed by her predecessor with Portman Entertainment, a London television production house that wanted to expand into feature films. The plan was to make a series of lower-budget films, light in tone. The series was given the awkward (and little-used) title Screenvisionz, and New Zealand On Air and Television New Zealand agreed to support it.

By sharing production costs with three other investors, the commission's money could be spread more widely, and it would not have to pay more than about 60 percent of the cost of each film. The first features were completed for as little as $1.5 million each. These were tight budgets, but they launched the careers of five young directors and producers. Four of the features also achieved Harley's aim of proving that New Zealand could make less serious films. In all of them, however, the comedic elements were underscored by an undeniable darkness of tone. By the fifth film this mood had become dominant.

First came *Via Satellite*, a comedy about a wildly dysfunctional family which comes together to watch the live broadcast of an Olympic swimming race in which one of the daughters is competing for a gold medal. Co-written with Greg McGee and directed by successful playwright Anthony McCarten, it was filmed in Wellington at the end of 1997 with a cast which included Danielle Cormack playing both the swimmer and her twin sister, Jodie Dorday and Rima Te Wiata as the other sisters, and Donna Akersten as their mother. Tim Balme played the husband of one of the sisters and Karl Urban the boyfriend of another.

When the film premiered at the Wellington Film Festival, the programme highlighted its 'cracking ensemble work from some of the finest men and women in New Zealand'. Columbia Tristar was, therefore, disappointed when its New Zealand release attracted fewer than 37,000 people.

It blamed lack of time, but I felt the company had underestimated the enormity of the task of launching a New Zealand film. For distributors of Hollywood films, each title reached New Zealand with a ready-made identity, boosted by months of exposure in international magazines, television shows and music charts, and supported by the lavish attention given to American releases and their stars. Each New Zealand film, however, started from scratch. Its very existence had to be established. And far from having the resources of Hollywood-propelled coverage, it had only the local media, which was often more interested in giving space and time to imported films. This was one of the reasons producers often wanted to launch their films offshore: if a film were acclaimed internationally, it might have a head start locally.

The second, and most popular, Screenvisionz film was a comedy-thriller about university students. *Scarfies* was shot in the university city of Dunedin with a script by the director, Robert Sarkies, and his brother Duncan. The film, by turns hilarious and brutal, portrayed a group of students discovering a marijuana-growing operation in the basement of their rented house, and conspiring to take over both it and its owner. The New Zealand release in 1999, astutely handled by independent distributor Kelly Rogers of Essential Films, grossed over $1.2 million and was seen by more than 160,000 people. The movie won six New Zealand film awards, including best film and best director, and sold to a dozen countries. Robert Sarkies went to Spain for the 80-print theatrical release and returned with a poster showing the film re-titled *Marijuana*.

Next up in the series was *Savage Honeymoon*, a comedy written by its director Mark Beesley. In it, a couple's second honeymoon is gatecrashed by a demanding mother-in-law and their two teenage children. *New Zealand Herald* reviewer Peter Calder found it broad and unsubtle, but it garnered five New Zealand film awards and was seen by 37,000 people.

This wasn't enough, though, to cover its considerable release costs – a fate which would also befall Michael Hurst's directing debut *Jubilee,* a gentle – too gentle – comedy of small-town life based on a book by Maori author Nepi Solomon. *Jubilee*'s leading man was Cliff Curtis, who returned home after having played supporting roles in six Hollywood features, including Martin Scorsese's *Bringing Out The Dead* and Michael Mann's *The Insider.* Curtis played the lovably disorganised Billy Williams, a procrastinator who is given the hopeless task of organising a school reunion. The role, which won him the year's best actor award, was a notable contrast to the villains he had played in *Desperate Remedies* and *Once Were Warriors* – performances which had led to still more villainous roles in Hollywood.

Fourth in the Screenvisionz series was Hamish Rothwell's *Stickmen*. Another mix of comedy and thrills, this lively film had a sexy cast including Robbie Magasiva, Scott Wills and Paolo Rotondo as best friends who risked their lives when they entered a pool competition run by a sinister criminal. The New Zealand cinema release in 2001 attracted 75,000 people, and the film won best director for Rothwell, best actor for Wills, and best script for writer Nick Ward. The film also sold to 25 countries, matching the sales record of *Via Satellite*. But a pre-sale to Universal Pictures International for Britain ended disappointingly: although the company bought all rights, it decided to release the film straight to video. The high costs of a cinema release were, it said, too risky for a film in which director and cast were unknown.

Comedy was harder to discern in *Snakeskin*, the last film in the series. Writer-director Gillian Ashurst and producer Vanessa Sheldrick cast *Heavenly Creatures'* Melanie Lynskey in their violent and briefly supernatural road movie. My doubts about the film were confirmed when *Snakeskin*'s New Zealand release attracted fewer than 30,000 people, the lowest attendance of the series.

No matter how many New Zealand films were screening at each market – two at Cannes in 1997 and 1998, and eight in 1999 – the commission's task remained the same: to maintain the interest created by each successful feature until the next success came along. We appeared to have succeeded. British critic Nick Roddick, who kept a close eye on us, wrote in *Sight and Sound* that we had maintained a sense of both continuity and buzz. As well as the annual markets, we had been active on many other fronts, particularly film festivals. In the mid '90s we had placed New Zealand feature films in 70 festivals within a year. The resulting sales had brought significant foreign exchange, and the Film Commission had won an export award.

By the end of the 1990s, things were looking good, and government enthusiasm for the film industry was on the rise once more. There had been increases in the commission's annual budget: grants from the government and the Lottery Board were now holding at around $10 million a year. And the five Screenvisionz features, for all their varying successes, had helped achieve everyone's aim of lifting the number of feature films being made in New Zealand by New Zealanders.

One young girl dared to confront the past
change the present
and determine the future

A film by Niki Caro

WHALE
RIDER

WINNER
TORONTO FILM FESTIVAL
PEOPLE'S CHOICE AWARD

Based on the novel by Witi Ihimaera

SIXTEEN

RIDING
THE WAVE

*T*HE BOOM TIMES HAD RETURNED. In 1999 a record eleven local feature films screened in New Zealand cinemas. At Mifed in October we advertised 12 new features – five we were screening and seven in post-production. Eight were from first-time directors. Film Commission investment was continuing to create many opportunities for both established and aspiring film-makers.

Such opportunities were, however, not always appreciated. In 1997 we hired publicist Anne Chamberlain to organise an anniversary dinner celebrating the first 20 years of productions financed by the commission. After a week of contacting film-makers, Anne phoned to report that some of them wanted to talk about what they thought was wrong with the commission, rather than about a celebration.

We asked Ian Mune, who had a distinguished and impressive all-round film career as an actor, writer and director, to be master of ceremonies at the dinner. The commission had paid a subsidy for an anniversary issue of the industry magazine *Onfilm* to be produced and distributed to guests. When we opened the magazine on the day of the dinner we were shocked to discover that Mune had contributed a barbed attack on the commission. Although acknowledging that its investment decisions had achieved 'a pretty good hit rate', Mune accused it of flailing uncertainty and Machiavellian secrecy, and proposed that in future the government appoint only film-makers to the board.

Peter Jackson had also been invited to contribute to the anniversary issue. We learned later that he had sent a critical article which *Onfilm* had decided not to publish. The article turned up the following month in *Metro* magazine, in a version described by the editor as legally revised. In it Peter called for total reinvention of the commission and complained about 'constant revisionism of policies'. Like Mune, he proposed a board consisting entirely of industry members. This would, he said, overcome film ignorance at board level, which he saw as one of the biggest problems. It was, he added, also time for fresh blood among the staff. We wondered whom he had targeted for slaughter.

Jackson repeated his criticism in an interview with the American trade paper *Variety*. He talked about an 'unpleasant experience' when the commission had withdrawn its investment from *Pink Frost*, a psychological thriller being developed by Harry Sinclair with Jackson as one of the producers. He would, he said, be meeting film industry colleagues to discuss the massive, sweeping changes that were needed.

Other producers, including Robin Scholes and John Barnett (then a board member for a second term) were more sanguine, and veteran producer John O'Shea could see both sides. 'The idea of fostering and encouraging independent production has triumphed,' he acknowledged. 'New Zealanders and many interested in films worldwide now know the names of Roger Donaldson, Geoff Murphy, Vincent Ward, Peter Jackson, Jane Campion, Gaylene Preston and Lee Tamahori.' But the commission, he felt, had lost its nerve. 'Television channels and overseas interests too easily hold the purse strings … Timidity rules.'

The issue of television having an undue influence had also been raised by Peter Jackson, who said the commission should stop investing in television co-ventures. Failing to foresee the cinema successes of two of the Screenvisionz features, he worried that the series was steering young film-makers into starting their careers with tele-features. He also criticised the low pay rates. He was right about the pay rates – yet he had started his career with a budget one-fifth the size of the ones he was criticising.

Debate about the commission and its policies would continue, on and off the record, for much of the year, at venues from Jackson's offices to Parliament Buildings. But choosing board members was the prerogative of the government, and politicians were not persuaded that an all-film-maker board was needed.

In the meantime, the commission continued to invest in New Zealand films – including a big one being made by one of its critics.

Ian Mune's sixth feature was his fifth in which the commission had been a major investor. Released 18 months after the controversial anniversary, *What Becomes Of The Broken Hearted?* was the sequel to *Once Were Warriors*, whose director Lee Tamahori was now working in Hollywood. It featured another powerful performance by Temuera Morrison as (now reformed) wife-beater Jake Heke.

The producer, Bill Gavin, chose an Australian company to sell the film overseas. He told me he would make sure it got higher advances than I had achieved for *Once Were Warriors*. In the event, income fell short of the *Warriors*' figures – unsurprisingly, as sequels rarely do more business than the original films.

Nevertheless, the film did very well in its home market. Following the same strategy as *Once Were Warriors*, it was released in New Zealand just days after its launch at Cannes. The result was ticket sales of $3.2 million and a New Zealand audience of nearly half a million. The ticket sales were the second best any New Zealand film had achieved at home, and the attendance the third best. The results were, however, less than half those of *Once Were Warriors*.

For Mune, the film became his biggest domestic success, not only at the box office but also at the New Zealand film awards, where it won nine prizes, including best director for Mune, best actor for Morrison and best screenplay for Alan Duff.

What Becomes Of The Broken Hearted? was equally successful in Australian cinemas, where it grossed more than A$3 million. But in some other countries where *Once Were Warriors* had been a cinema success, the distributors decided to send the sequel direct to video. This was a particular frustration in the United Kingdom. Because *Once Were Warriors* had done great business for one of its competitors, Polygram had made a large financial commitment to get the sequel, but when it saw the completed film it had second thoughts about a theatrical release.

The company's German office reached the same conclusion. A distributor told me he felt the film looked like a Hollywood gang movie but without the American accents, and lacked the unique New Zealand dimension that had made *Once Were Warriors* such a success. This opinion was reflected in the total number of international sales: more than 100 for *Once Were Warriors*, about 20 for the sequel.

Success for New Zealand movies would reach new heights with a much gentler film. Niki Caro's talent as a director had first been spotted with her short film *Sure To Rise* at Cannes in 1994. Four years later her first feature *Memory and Desire*, a tragic tale of a Japanese bride whose husband drowns during their New Zealand honeymoon, hadn't become the art-house success we had hoped, but had earned a place in Critics Week at Cannes and been voted best film at the New Zealand film awards. It had been sold from London by Wendy Palmer, who had handled *The Piano*. Starting with pre-sales, Wendy licensed the film almost everywhere, except the United States and the United Kingdom where it was never released.

In 2002 Niki Caro completed her second feature, *Whale Rider*, with astonishing results. *Whale Rider* was based on a novel by acclaimed Maori author Witi Ihimaera, in which a young girl shows all the traits to be her tribe's new leader but is overlooked by her grandfather, who is determined to find a male worthy of the role. Co-producer

John Barnett had begun developing the project ten years earlier in partnership with Murray Newey. After Barnett left to join South Pacific Pictures, Newey didn't renew his option on the rights and Barnett picked them up. Caro wrote a script – replacing an earlier one by Ian Mune – and filming was completed in December 2001.

From the world premiere at the Toronto Film Festival, where *Whale Rider* was voted most popular film, Witi Ihimaera commented: 'In a world where film-goers think they've seen everything, *Whale Rider* and all our films about Maori take them to a place they've never been, a place where they can make special discoveries …'

Almost 850,000 New Zealanders saw *Whale Rider*; the box-office of $6.4 million was second only to the record set nine years before by *Once Were Warriors*. In Australia the box office hit A$8 million. In the United States it was more than US$21 million, seven times what any other New Zealand film – apart from *The Piano,* not officially a New Zealand film, and *The Lord Of The Rings* trilogy – had achieved.

The success first of *Once Were Warriors* and then of *Whale Rider* sent a clear message to film-makers: New Zealand's two top-performing movies had told universal stories, but with characters and backgrounds that could not have come from anywhere else. The same point had been made by *New York Times* critic Andrew Johnston, reviewing a season of eight New Zealand features in 2001. 'Save for local accents and references,' he wrote, 'some of the films were nearly indistinguishable from many American films. The best were those that used local culture and folklore, not as mere window-dressing but to add resonance and depth to established genres.' Too many film-makers had tried to imitate Hollywood, without noticing that the market was fully supplied with the real thing.

Twenty-five years of New Zealand films being released at home and overseas has given us plenty of information about success and failure. The list of disappointments has included almost everything that has claimed to be 'international', unless backed by an international studio to guarantee a release.

The question of what films to make is further complicated by the fact that overseas our films (like other 'foreign' films) are screened at art-house cinemas, whereas in New Zealand film-makers are encouraged to aim at multiplex audiences. After the 1980s' successes of Donaldson and Murphy, efforts at making populist films for New Zealand audiences often produced features that failed to get a theatrical release elsewhere, because they had no attraction for international art-house audiences. This dilemma has never been resolved.

Late in 1999 I was invited to Bangkok to present a paper on film marketing to a conference of Asian producers at the Asia Pacific Film Festival. As the sole New Zealand representative, I was rewarded with a limousine, a driver, a minder and a motorcycle escort. In the early 1980s Film Commission chair Bill Sheat had regularly represented New Zealand at this conference, and in 1995 we had hosted it in Auckland. Participation offered the chance of making connections with wealthy Asian producers, but New Zealand producers seemed to prefer to seek production money in Europe or North America, rather than from countries closer to home.

The search for offshore production money could frequently be disappointing. John O'Shea had twice persuaded German financiers to invest in his films, but after the failure of *Te Rua* in 1991 he couldn't find backing for other projects. Potential investors were no doubt further discouraged by the fact that he was by then in his seventies. Nevertheless, John continued to develop projects and tell everyone about them. His most visionary was *Sutherland Falls*, the story of a nineteenth-century South Island pioneer, which he wrote with the aim of having Scottish actor Billy Connolly play the lead. But O'Shea never found the offshore finance for this project to go ahead.

Murray Newey faced similar difficulties after completing his two co-productions with Canada. In 1997 he was one of nine New Zealand producers who went to the American Film Market seeking investment money. Sixteen New Zealand projects were taken to the market. None went into production. Newey was hoping to make *Lake Of Lost Souls*, an appealingly scary horror film written by Michael Heath for director David Blyth. A year later, with lack of production just one of his anxieties, Murray committed suicide. His many film-making friends attended his funeral, shocked by the loss of one of the industry's most genial entrepreneurs.

For producer Trevor Haysom it would take more than a decade to get a second feature on the road after the disappointing results of his first production, *User Friendly*. He stayed active by associate-producing two features and producing a variety of short films, until finalising finance for *In My Father's Den* as an official New Zealand–United Kingdom co-production. The movie, based on a 1972 novel by award-winning New Zealand author Maurice Gee, told the story of a war-weary photographer returning to his home town and getting caught up in the disappearance of a teenage girl, played in a bravura performance by teenager Emily Barclay. Written and directed by Brad McGann, it became a substantial success in 2004, exceeding $1.5 million at the local box office, and winning acclaim and awards at festivals in Canada and Spain.

After the collapse of Mirage, Larry Parr didn't produce another feature film for nine years. In 1997 he returned with a new company, Kahukura Productions, whose first film, *Saving Grace*, was also a first for its director Costa Botes. For the film's world premiere we offered it to a festival in Valladolid, a Spanish city which seemed to be filled with cathedrals and palaces and was famous as the home of Miguel de Cervantes, author of *Don Quixote*. The festival was also presenting a retrospective of New Zealand cinema, and we felt Spanish sensibilities would be well matched with the story of a homeless teenager (played by Kirsty Hamilton) who falls in love with an unemployed carpenter (played by Maori actor Jim Moriarty) who then tells her that he is Jesus Christ.

We hadn't anticipated the intensity with which the Spanish would receive a film on such a subject. At a late-night press conference after the official screening, Spanish critics questioned Botes lengthily and in considerable detail about the ambiguity of the film's ending. Was the carpenter Jesus or wasn't he? When Botes returned home, he and Parr decided to shoot a different, more clear-cut ending.

One of the few countries to buy the film was Canada, where I made a deal with Jim Murphy, who had changed distribution companies since I had sold him *Once Were Warriors*. Jim, who had discovered the film at one of our market screenings in Los Angeles, preferred the original ending. I stepped back from the debate and watched a lengthy e-mail correspondence between distributor and director about which version would be released in Canada. After his Spanish experience, Botes understandably wanted the new ending. Murphy insisted on the original. He got what he wanted, refusing to sign the contract otherwise. But when his Canadian release began at Easter 1999, after a premiere at the Montreal Film Festival, Canadian audiences and critics did not share his enthusiasm.

When I organised the market screenings of *The Irrefutable Truth About Demons* and *The Price Of Milk* at Cannes in 2000, it felt more like business as usual than a farewell appearance. But I had decided several years earlier that it was time to end the unrelenting round of marketing and selling and negotiating and contracting. The pleasure of so much international travel was beginning to pall. I didn't want to queue for any more long-distance flights or check in to any more international hotels. I wanted to stay in one place for 12 months, and to experience all four seasons at home.

So I didn't have any regrets that this would be my last working visit to Cannes – except when it came time to farewell Monique Malard and the other local staff who had been a part of my life every May for 21 years. After the festival, I went with Monique and her husband to our favourite Vietnamese restaurant. We stayed late into the night, talking about the many experiences we had shared and the many New Zealand film-makers we had worked with.

The Film Commission, with distributor-turned-exhibitor Barrie Everard as its fifth chair, farewelled me at Cannes – and then again ten months later at an industry party in Wellington. A movie of my life, directed by Robert Sarkies, showed a mélange of scenes from Cannes over a soundtrack of Tony Bennett singing 'The Good Life'. Gaylene Preston, Glenn Standring, Jane Campion and Vincent Ward were among those who spoke.

Vincent, home after almost a decade in Los Angeles, was beginning a prolonged development period for his fifth feature, *River Queen*. Although he was at the party, he had chosen to pre-record his speech. It was followed by a videotaped message from Peter Jackson talking amiably on one of his *Lord Of The Rings* locations about his excitement when I started to sell *Bad Taste*. He offered a hobbit house for my retirement.

Peter's shaggy informality had become synonymous with the highest achievements of New Zealand film-making. A few months earlier, to mark the end of the 15-month shoot of *The Lord Of The Rings,* the soon-to-be-world-famous director had thrown a spectacular party in an immense, neglected Victorian woolshed on the Wellington waterfront. It had featured entertainment and decor on a scale probably never seen before in New Zealand. The next morning I bumped into Peter, not yet a recognised figure, walking along Willis Street holding a cluster of shopping bags containing Christmas gifts.

The day after my last market in Milan, I took a train to Rome to visit Guglielmo Biraghi. He was in hospital in the last weeks of a fatal illness. Our good friend Jeannine Seawell had died two years earlier. Both had suffered lung cancer after years of heavy smoking. At Jeannine's last Cannes Film Festival she had chosen not to tell anyone she was ill. We had set a date for a dinner in Paris. It never happened.

A few years before, we had asked each other how long we would stay in the film business. Although she had been selling films for almost 30 years, Jeannine had said she wanted to stay in the job until she died. I felt differently. I would choose a time to leave, I told her, because too much of the negotiating was coming to seem predictable, and too many of the routines of festivals and markets wearisome. It was a decision I would sometimes reflect on, but never regret.

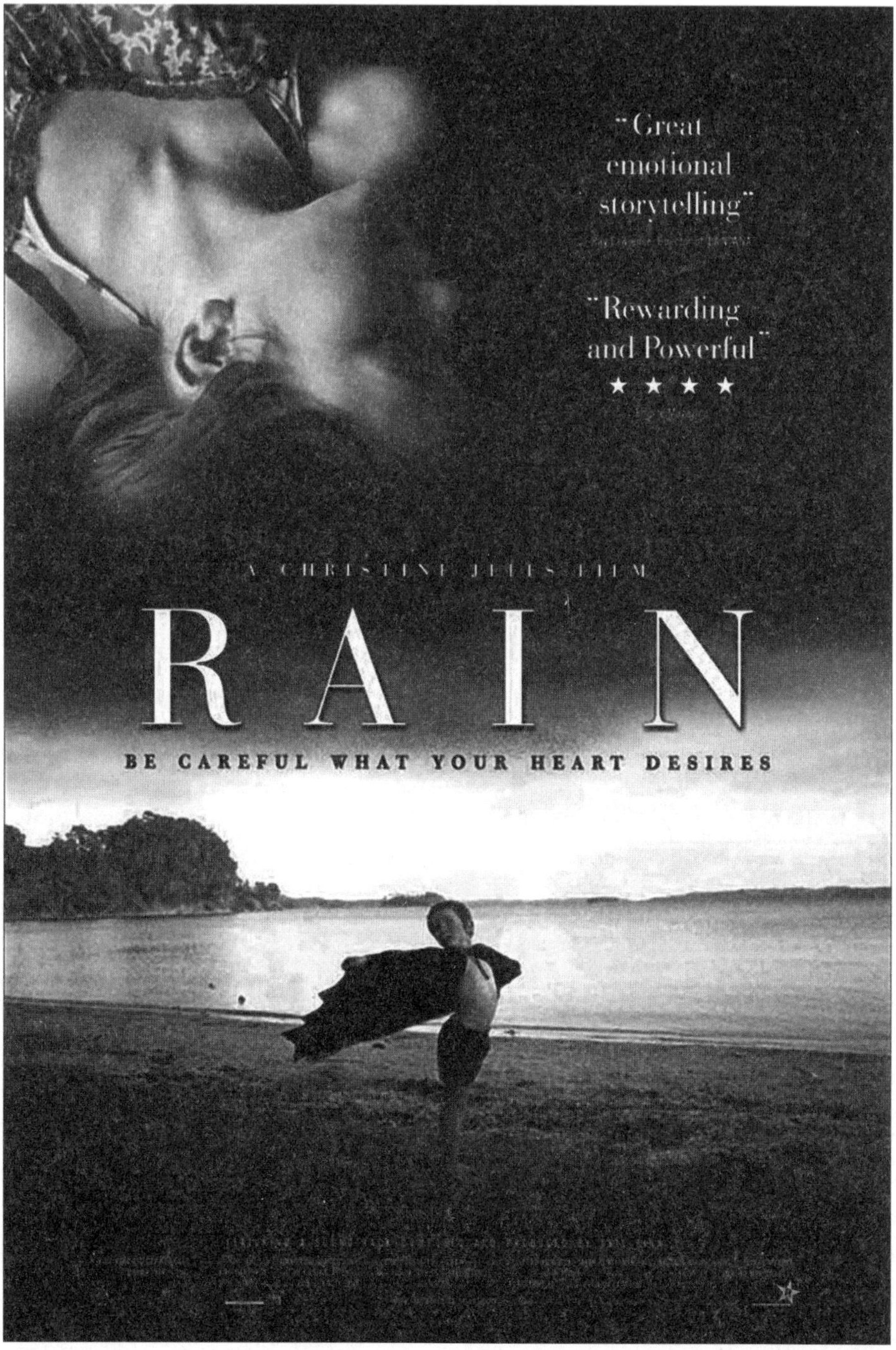

"Great emotional storytelling"
"Rewarding and Powerful"
★ ★ ★ ★
A CHRISTINE JEFFS FILM
RAIN
BE CAREFUL WHAT YOUR HEART DESIRES

POSTSCRIPT

*I*N MAY 2001 I STAYED AT HOME in Wellington while my successor, Kathleen Drumm, flew to Cannes, where she launched Christine Jeffs' movie *Rain* with screenings in the Directors Fortnight and sales to many countries, including the United States and France. This was my first May in New Zealand in 21 years. It was an Indian summer.

Two months later John O'Shea died at the age of 81. The memorial service for the father of modern New Zealand film-making was held in the Embassy Theatre, a 1920s picture palace in the heart of the city. Extracts from John's films were screened, and a score of film-makers remembered how he had helped them enter the film industry by giving them work at Pacific Films.

The Embassy was a fitting venue for such an occasion. Five years earlier, when it seemed nobody in the film business wanted the run-down theatre, Bill Sheat had had the bright idea of forming a not-for-profit trust to save it. Bill Gosden, David Carson-Parker and I were among the group whom he invited to become the trust's foundation members, joined in the second year by Peter Jackson and Fran Walsh. We were all passionately committed to keeping the Embassy in daily use as a cinema and restoring it to its neo-classical elegance.

One of my personal motivations was to provide a permanent home for the Wellington Film Festival. The festival had outgrown the nearby Paramount Theatre and moved to the Embassy in 1983. Despite the Embassy's shabby condition, it was a perfect festival cinema. The largest in Wellington, it had 850 seats, perfect sight-lines, and a huge screen which had been installed in front of the original proscenium arch in 1960. Even before restoration started, the Embassy was a magnificent venue in which the festival kept on growing, in spite of the threadbare carpets and torn curtains.

It had taken three years before the trust had been able to find the financial guarantee necessary to raise a loan and buy the building. The purchase may have saved the Embassy from being converted into a supermarket or apartments, but the challenge had barely started. Fund-raising was to continue for six more years. Embassy Trust meetings, held most weeks, were often marked by emotional and sometimes acrimonious debate as we struggled to deal with the ever-increasing cost of restoring and strengthening the beautiful old building, and the elusive task of raising the money.

With Peter Jackson as a trust member, the cinema became the venue for the New Zealand premieres of the first two films in *The Lord Of The Rings* trilogy. Then in 2003, with the final restoration of the auditorium completed by only days, the grand old theatre hosted the flawlessly successful world premiere of *The Return Of The King,* and the film festival acquired a magnificent venue. All the effort seemed suddenly worthwhile.

The 30th Wellington Film Festival, which took place only a few weeks after John O'Shea's death, was the last at which Jonathan Dennis would occupy his regular seat in the front row. The man whose determination and persistence had achieved the founding of the New Zealand Film Archive, and who had been its first director, would die of cancer the following year aged 48. After leaving the archive in 1990, Jonathan had spent a decade editing books, including the important *Film in Aotearoa New Zealand*, making a documentary about a veteran cinematographer, and presenting his weekly *Film Show* on Radio New Zealand. It sometimes seemed everyone with an interest in movies tuned in to hear his acerbic opinions of films and film-makers.

Before he died, he specified that his memorial service be held in the Paramount, the film festival's original home. At the funeral, friends and family stood alongside his multi-coloured coffin as they spoke about his quiet achievements. Samples were shown of his favourite films saved by the Film Archive, and clips were played from his radio show, culminating in a pithy comment about Gwyneth Paltrow, never one of his favourite actresses. We walked out of the theatre laughing and crying at the same time.

Two close colleagues who had left the commission before me stayed involved with film. Ruth Jeffery became founding director of the New Zealand Film and Television School. Catherine Fitzgerald, first a board member and then policy director, began a producing career. Her first short film *Turangawaewae*, directed by Peter Burger, earned selection in Critics Week at Cannes; her second, *Two Cars, One Night*, directed by Taika Waititi and co-produced with Ainslie Gardiner, won nine international awards, including best short film in the Berlin Film Festival's panorama.

The first two chairs also kept up their involvement in building and nurturing the industry. Bill Sheat became chair of a trust overseeing New Zealand film festivals, which had proliferated to a dozen cities. David Gascoigne was appointed chair of the Film Production Fund, set up in June 2000 to administer an extra $22 million of government financing. Niki Caro's *Whale Rider* and Gaylene Preston's *Perfect Strangers* – starring Sam Neill in his first New Zealand feature since *The Piano* – were the first productions it supported. They were followed by Glenn Standring's *Perfect Creature* and Vincent Ward's *River Queen,* both co-productions which also raised substantial money from the United Kingdom, and Roger Donaldson's *The World's Fastest Indian* which also had investment from Japan.

Funding, or the lack of it, continued to be an obstacle for many film-makers, although there were some imaginative solutions. At Kahukura, Larry Parr initiated a group of four 'no budget' features. In fact the brand was misleading: the Film Commission had agreed to provide finance, although at a controversially low level, for both production and (conditionally) post-production. After the unfortunately titled *Hopeless*, the next to be released was also the most striking and original: Vanessa Alexander's *Magik And Rose*, set in Hokitika on the west coast of the South Island, and described by *Variety* as disarmingly sweet. It was seen by more than 26,000 New Zealanders and ran for two months in the southern cities of Christchurch and Dunedin.

Completion of the other three features was delayed when Kahukura went into voluntary liquidation at the end of May 2002 while Parr was directing *Fracture* (another film based on a Maurice Gee novel). Six and a half months later Peter Jackson launched a fresh attack on the commission. The wording of his single-spaced two-page press release, sent to me from the office of Wellington lawyer Michael Stephens, seemed out of character with the man I had known for 15 years – always conciliatory and constructive, willing to find solutions to the most difficult problems.

Peter was angry that Kahukura's collapse had left unpaid bills at the Film Unit, which he now owned. There was a debt of $180,000 for post-production costs on *Kombi Nation*, one of the 'no budget' series, and a further $80,000 for *Fracture*. His statement was dated one day before the New Zealand premiere of *The Two Towers*, the second film in *The Lord Of The Rings* trilogy. He announced that the commission would not be welcome at the event.

The commission had in fact been working behind the scenes to fix the complex problems caused by the Kahukura liquidation, with former chairman Alan Sorrell as a legal adviser. But Jackson's view was that film crews and suppliers – who had also been left with unpaid accounts – had been 'abused and vilified by these self-serving bureaucrats'. The issues between Jackson and the commission were resolved the following March, when a statement agreed by both sides said 'satisfactory commercial settlements' had been reached with the Film Unit and the liquidator. Both settlements were subject to confidentiality agreements.

After the sorting out, two of the uncompleted 'no budget' features were able to be finished. *Kombi Nation*, a raucous black comedy by Grant Lahood about a European holiday from hell, was released in August 2003 and seen by 21,000 New Zealanders. Stuart McKenzie's psychological thriller *For Good* saw the light of day in 2004. It failed to find a local audience but won a prize for its leading actress, Michelle Langstone, at a film festival in St Tropez.

With the collapse of Kahukura, the established industry seemed to lose faith in low-budget movie-making. Producers searched for larger amounts of money, in the hope this would provide a winning 'cinematic' element. I remained sceptical. I hoped the industry hadn't forgotten it was the idea, the script and – as the great Italian director Federico Fellini had said in one of his last interviews – the vitality of the storytelling that were all-important.

Meanwhile, a young generation of film-makers, less bothered by small budgets, looked instead to low-cost digital technology to make their movies. In 2003 a new wave of these feature films – seven in all – appeared. All had been made with miniscule budgets. The organisers of the year's film awards created a new section for digital features, and all four categories were won by Gregory King's *Christmas*, a bleakly engrossing story of a dysfunctional family, which the director had filmed in his family home in Whangarei.

The other digital features included German-born director Florian Habicht's quirky *Woodenhead*, which the film-maker summarised as a story about a shy dump-hand who has to escort a mute princess to her wedding; Campbell Walker's drama of phone harassment, *Why Can't I Stop This Uncontrollable Dancing*; and Alexander Greenhough's *I Think I'm Going*, described by Bill Gosden as 'scrutinising with insight and irony the listless love lives and insecure domestic arrangements of a group of Wellington 20-somethings'. The cinema of unease wouldn't die.

A trend for expatriate directors to return to New Zealand and make movies had started in 2000, when Sam Pillsbury came back from Los Angeles to direct *Crooked Earth* for producer Robin Scholes. One year later, Geoff Murphy completed his first New Zealand feature since *Never Say Die*. After 12 years in Hollywood, making action movies such as *Freejack* and *Young Guns II*, he looked to the past and directed a documentary about the glory days of Blerta, the anarchic travelling music show which had been loosely headed up by his sometime friend and brother-in-law Bruno Lawrence.

Then came Murphy's seventeenth feature, a conspiracy thriller entitled *Spooked*, which he wrote as well as directed. *Spooked* at first failed to attract Film Commission funding, but then the commission changed its mind and provided the finance. The film bombed at the local box office. Audiences either stayed away or, worse, walked out. The Murphy magic seemed to have vanished.

Vincent Ward would complete *River Queen* – a story of New Zealand's 1860s land wars, written by the director himself with Toa Fraser – in 2005 after a winter shoot on the Whanganui River, with Cliff Curtis and Temuera Morrison among the local cast, and British actress Samantha Morton making headlines when she became ill and production was delayed.

Roger Donaldson also took a plane home, turning down what would have been his twelfth Hollywood feature to make a New Zealand film he had been dreaming about since the '70s. *The World's Fastest Indian* would tell the astonishing story of a man from Southland setting a land-speed record at Bonneville Salt Flats in Utah, with British star Anthony Hopkins in the leading role. Like *River Queen*, the film was being readied for release as this book went to print.

Rena Owen came home on several unpublicised visits. Now a producer and living in Los Angeles, she worked with *Once Were Warriors*' scriptwriter Riwia Brown to create a landmark project – *Behind the Tattooed Face* – the first feature film to be set in New Zealand in the 1700s, before European colonisation. Based on a novel by Heretaunga Pat Baker, it will have an all-Maori cast.

Judith McCann came home too, for a second time, returning from Canada to head Film New Zealand, a body created by the industry to assist – and, more controversially, encourage – offshore productions to use New Zealand locations. With its focus broadening to include the promotion of talent and facilities, Film New Zealand now receives all its financial support from the government, with a group of public servants emerging to boost its promotional efforts. If some of these new arrivals believe no one knew anything about the New Zealand film industry before *The Lord Of The Rings*, then this book will be part of their learning curve.

After 25 years of film-making, local productions have become a source of national pride for New Zealanders – and not only because of the unparalleled successes of

Peter Jackson and *The Lord Of The Rings*. They've also become a unique cultural resource, ready to be studied and enjoyed by generations who didn't see them when they were first released. By the end of 2004 the Film Commission was administering the earnings of more than 130 feature films and 390 short films. Its annual budget had been almost doubled by prime minister Helen Clark's Labour government, which had also set up a generous (some said over-generous) grants scheme to encourage big-budget productions, primarily from offshore, to use New Zealand locations.

The Film Archive's collection had grown too, until it exceeded a million items, with more than 20,000 titles available for viewing. And the various worlds of New Zealand film were intermingled. The commission, which had taken the first steps to form the archive, became its tenant for some years, while the archive became the guardian of prints and negatives for the films financed by the commission. The film societies presented annual programmes of these films, and the film festivals looked to the archive for some of their retrospective programming.

The Embassy Theatre became not only the home of the Wellington Film Festival but also a focus for film industry premieres and special events. In October 2004 it was packed for the premiere of *Tama Tu*, a marvellously entertaining and good-humoured short film about the Maori Battalion, directed by Taika Waititi and produced by Cliff Curtis and Ainsley Gardiner, three of the new generation of Maori film-makers whose commitment to 'telling our stories' echoes the ambitions and words of the generation which campaigned for a film industry 30 years earlier.

The nagging question still remains: has the flowering of New Zealand feature films – more than 200 in the 25 years to 2004, compared with only 15 in the previous 25 years – enabled us to escape the domination of Hollywood with which I grew up? Certainly local media seem happy to stay dominated. Fed by a ceaseless diet of films and film gossip from the United States, they can appear to have stars in their eyes concerning anything and anyone connected with Hollywood, while being reluctant to give equal space or time to locally produced films.

And when space *is* found, the bar is set high: a New Zealand film must be a hit at the box office or critics rush to brand it a failure, with little attempt to identify individual effort or achievement. Yet it is a truism that only a small proportion of any film industry's total production can set box-office records. Given that the number of films produced in New Zealand since film-making began is less than the total produced by Hollywood in a year, the success rate is extraordinary. Most phenomenal of all is Peter Jackson who, while continuing to live and work in New Zealand, has become one of the most influential people in Hollywood.

If New Zealanders can never disentangle ourselves from Hollywood, we can be proud that it's no longer a one-way relationship. We now have our own directors telling stories with local resonance and recognition, in films which have won international acclaim.

Best of all, our film-going experiences are no longer solely from somewhere else. Our most memorable movie characters now include plenty with New Zealand accents – the three Janets of *An Angel At My Table*, Geoff and Firpo on Takapuna Beach, the Tainuia Kid, the denizens and dogs of *Footrot Flats*, the *Bad Taste* boys, Lionel and his mum, Jake and Beth, Pauline and Juliet, Paikea and her grandfather, and even a rapidly disintegrating, oddly endearing little yellow Mini.

FILMOGRAPHY

*I*N THE 25 YEARS SINCE I STARTED keeping track of New Zealand feature films, one thing has been certain: whenever the list seemed to be perfect, someone would point out a title that had been overlooked, or was described inaccurately. The list which follows is the most complete and the most substantial so far. But experience tells me to beware of claiming perfection.

To qualify as a feature, each title must have a duration of 70 minutes or more, which is the reason films such as *A State Of Siege* and *Forgotten Silver* aren't included.

The list includes made-for-television movies, and documentaries which had a brief theatrical release (usually in film festivals) before being telecast.

If a feature wasn't released with 35-millimetre prints, the format is in brackets after the title (16-millimetre, digital, DVD, video). Titles are listed by the year of first public screening – which includes film festivals – at home or overseas. A few titles don't seem to have been released at all, and these are so identified.

The 254 features which follow were all created, filmed and completed in New Zealand. Well, almost all of them. I have allowed one or two exceptions to the rule – most obviously *The Piano*, which is officially an Australian feature (because it was produced by an Australian) but which is so identified with New Zealand and with its New Zealand director that it would be illogical to omit it.

Two other exceptions are John Reid's *Leave All Fair* and Larry Parr's *A Soldier's Tale*. Both were filmed in France, but both began their lives in New Zealand, were produced and directed by New Zealanders and completed here as well.

Purists will identify a few co-productions which completed post-production overseas, although always under the control of their New Zealand director. These qualify for the New Zealand list not only because of their content but also because of the dual nationality conferred by official co-production agreements. Official dual nationality is also the reason a film shot in England by a New Zealand director with a British cast is on the list.

The 49 features in the 'locations' section were filmed in New Zealand, but created and completed in other countries, and (except for the silent movies) couldn't by any stretch of the imagination be described as New Zealand films. There's nothing of New Zealand in *Vertical Limit* or *The Last Samurai* except the scenery, and even this is pretending to be somewhere else.

On the other hand, Peter Jackson's record-breaking trilogy *The Lord Of The Rings* easily qualifies as New Zealand-made. All three of the movies have official New Zealand nationality and all three were created, filmed and completed in New Zealand.

NEW ZEALAND FEATURE FILMS SINCE 1914

1914

Hinemoa
Director/producer: George Tarr
Photographer: Charles Newham

1916

The Test
Produced by and starring: Rawdon Blandford

1922

My Lady Of The Cave
Director: Rudall Hayward

Ten Thousand Miles In The Southern Cross
Director: George Tarr

The Birth Of New Zealand
Director: Harrington Reynolds

1923

The Romance Of Sleepy Hollow
Director: Henry Makepeace

1924

Venus Of The South Seas
Director: James R. Sullivan

1925

Glorious New Zealand *(documentary)*
Government Publicity Office

Rewi's Last Stand
Director: Rudall Hayward

The Adventures Of Algy
Director: Beaumont Smith

1927

Carbine's Heritage
Director: Edwin (Ted) Coubray

The Romance Of Hine-Moa
Director: Gustav Pauli

The Te Kooti Trail
Director: Rudall Hayward

1928

The Bush Cinderella
Director: Rudall Hayward

1934

Romantic New Zealand: Land Of The Long White Cloud *(documentary)*
Filmcraft

1935

Down On The Farm
Directors: Lee Hill and Stewart Pitt
New Zealand's first sound feature film

1936

On The Friendly Road
Director: Rudall Hayward

Phar Lap's Son?
Director: A.L. Lewis

The Wagon And The Star
Director, producer and screenplay: J.J.W. Pollard

1940

Rewi's Last Stand (UK title ***The Last Stand***)
Director, producer and screenplay:
Rudall Hayward
Photography: Rudall Hayward, Edwin Coubray,
Jack McCarthy
Editor: Rudall Hayward
With Leo Pilcher and Ramai Te Miha (later Ramai
Hayward)

1950

1950 British Empire Games (documentary)
National Film Unit – no individual credits

1952

Broken Barrier
Directors and producers: John O'Shea,
Roger Mirams
Screenplay: John O'Shea
Director of photography: Roger Mirams
With Terence Bayler, Kay Ngarimu, Mira Hape,
Bill Merito, George Ormond, Lily Te Nahu,
Dorothy Tansley, F.W. French

1964

Runaway
Director and producer: John O'Shea
Screenplay: John Graham, John O'Shea
Director of photography: Tony Williams
Editor: John O'Shea
With Colin Broadley, Nadja Regin, Deirdre
McCarron, Selwyn Muru, Barry Crump,
Kiri Te Kanawa, Tanya Binning

1966

Don't Let It Get You
Director and producer: John O'Shea
Screenplay: Joseph Musaphia

Director of photography: Tony Williams
Editor: John O'Shea
With Howard Morrison, Carmen Duncan,
Gary Wallace, Alma Woods, Ernie Leonard,
Normie Rowe, Kiri Te Kanawa, Lew Pryme,
Rim D Paul, Herma Keil, Eliza Keil, Quin Tikis,
Tanya Binning, Harry Lavington

1972

To Love A Maori (16-millimetre)
Directors and producers: Rudall and
Ramai Hayward
Screenplay: Rudall and Ramai Hayward and
Diane Francis
Photography: Alton Francis
Editors: Rudall Hayward, Alton Francis
With Val Irwin, Marie Searell

1974

Games '74 (documentary)
Directors: John King, Sam Pillsbury,
Paul Maunder, Arthur Everard
Producers: David Fowler, Lance Connolly
– National Film Unit

1975

Landfall (16-millimetre telemovie)
Director and screenplay: Paul Maunder
Producer: David Fowler
Director of photography: Linton Diggle
Editors: Paul Maunder, Maxine Schurr
With Denise Maunder, John Anderson,
Sam Neill, Gael Anderson, Michael Haigh,
Pat Evison, Jonathan Dennis

Test Pictures (16-millimetre)
No director credited
Photography and editing: Geoffrey Steven
Screenplay: Denis Taylor

*With Denis Taylor, Lee Feltham, Francis Halpin,
Moira Turner, Geoff Barlow, Dora Warren,
Barbara Saipe, Mark Elmore, Mike Fitzgerald*

1976

The God Boy *(16-millimetre telemovie)*
Director and producer: Murray Reece
Screenplay: Ian Mune; based on the novel by
Ian Cross
Director of photography: Allen Guilford
Camera operator: Alun Bollinger
Editor: Simon Reece
*With Jamie Higgins, Maria Craig, Graeme Tetley,
Judie Douglass, Ivan Beavis, Bernard Kearns,
Yvonne Lawley, Dorothy McKegg, Sandra Reid*

1977

Off The Edge *(documentary)*
Academy Award nomination, best documentary
Director and producer: Mike Firth
Screenplay: Molly Gregory
Director of photography: Mike Firth
Editor: Michael Economou
With Jeff Campbell, Blair Trenholme

Sleeping Dogs
Director and producer: Roger Donaldson
Screenplay: Ian Mune, Arthur Baysting; based on
C.K. Stead's *Smith's Dream*
Director of photography: Michael Seresin
Editor: Ian John
*With Sam Neill, Ian Mune, Ian Watkin, Clyde Scott,
Nevan Rowe, Donna Akersten, Warren Oates,
Don Selwyn, Davina Whitehouse*

Wild Man
Director: Geoff Murphy
Producers: Bruno Lawrence, Roy Murphy
Executive producer: John Barnett
Screenplay: Bruno Lawrence, Geoff Murphy,
Martyn Sanderson, Ian Watkin

Director of photography: Alun Bollinger
*With Bruno Lawrence, Ian Watkin, Tony Barry,
Martyn Sanderson, Bill Stalker, Patrick Bleakley,
Val Murphy*

1978

Angel Mine
Director and screenplay: David Blyth
Producers: David Blyth, Warren Sellers
Director of photography: John Earnshaw
Editor: Philip Howe
*With Derek Ward, Jennifer Radford,
Myra de Groot, Mike Wilson*

Colour Scheme *(16-millimetre telemovie in the
Ngaio Marsh Theatre series)*
Director: Peter Sharp

Died In The Wool *(16-millimetre telemovie in the
Ngaio Marsh Theatre series)*
Director: Brian McDuffie

Skin Deep
Director: Geoff Steven
Producer: John Maynard
Director of photography: Leon Narbey
Editor: Simon Sedgley
Screenplay: Piers Davies, Roger Horrocks
*With Ken Blackburn, Deryn Cooper, Alan Jervis,
Grant Tilly, Glenis Levestam*

Solo
Director: Tony Williams
Producers: David Hannay, Tony Williams
Executive producers: Bill Sheat, John Sturzaker
Screenplay: Tony Williams, Martyn Sanderson
Director of photography: John Blick
Editor: Tony Williams
*With Martyn Sanderson, Lisa Peers, Vincent Gill,
Perry Armstrong, Davina Whitehouse,
Maxwell Fernie, Frances Edmond,
Veronica Lawrence, Val Murphy*

1979

Middle Age Spread (16-millimetre)
Director: John Reid
Producer: John Barnett
Screenplay: Keith Aberdein; based on the play by Roger Hall
Lighting cameraman: Alun Bollinger
Editor: Michael Horton
With Grant Tilly, Dorothy McKegg, Bridget Armstrong, Donna Akersten, Peter Sumner, Bevan Wilson, Ian Watkin, Yvonne Lawley

Sons For The Return Home
Director: Paul Maunder
Screenplay: Paul Maunder; based on the novel by Albert Wendt
Executive producer: Don Blakeney
Director of photography: Alun Bollinger
Editor: Christine Lancaster
With Uelese Petaia, Fiona Lindsay, Moira Walker, Lani Tupu, Anne Flannery, Alan Jervis, Sean Duffy, Tony Groser

1980

Beyond Reasonable Doubt
Director: John Laing
Producer: John Barnett
Screenplay: David Yallop; based on his book of the same name
Director of photography: Alun Bollinger
Editor: Michael Horton
With David Hemmings, John Hargreaves, Tony Barry, Martyn Sanderson, Grant Tilly, Diana Rowan, Ian Watkin, Terence Cooper, Bruno Lawrence

Goodbye Pork Pie
Director: Geoff Murphy
Producers: Geoff Murphy, Nigel Hutchinson
Screenplay: Geoff Murphy, Ian Mune
Director of photography: Alun Bollinger

Editor: Michael Horton
With Kelly Johnson, Tony Barry, Shirley Gruar, Claire Oberman, Bruno Lawrence, John Bach, Frances Edmond

Nambassa Festival (16-millimetre documentary)
Director: Philip Howe
Producers: Dale Farnsworth, Peter Terry

Squeeze (16-millimetre)
Director, producer, screenplay: Richard Turner
Director of photography: Ian Paul
Editor: Jamie Selkirk
With Robert Shannon, Paul Eady, Donna Akersten, Martyn Sanderson

1981

Pictures
Director: Michael Black
Producer: John O'Shea
Screenplay: Robert Lord and John O'Shea; from an idea by Michael Black
Director of photography: Rory O'Shea
Editor: John Kiley
With Kevin J. Wilson, Peter Vere-Jones, Helen Moulder, Elizabeth Coulter, Terence Bayler, Matiu Mareikura, Ken Blackburn

Race For The Yankee Zephyr
Director: David Hemmings
Producers: Antony I. Ginnane, John Barnett, David Hemmings
Screenplay: Everett de Roche
Director of photography: Vincent Monton
Editor: John Laing

Smash Palace
Director, producer and screenplay: Roger Donaldson
Associate producer: Larry Parr
Photographed by: Graeme Cowley
Editor: Michael Horton

With Bruno Lawrence, Anna Jemison,
Greer Robson, Keith Aberdein, Desmond Kelly,
Sean Duffy, Margaret Umbers
With Ken Wahl, Lesley Ann Warren, Donald
Pleasence, George Peppard, Bruno Lawrence,
Grant Tilly, Robert Bruce

Wildcat *(16-millimetre documentary)*
Directed, produced, written and filmed by
Rod Prosser, Russell Campbell, Alister Barry

1982

Battletruck
Director: Harley Cokliss
Producers: Lloyd Philips, Rob Whitehouse
Screenplay: Irving Austin, Harley Cokliss,
John Beech
Director of photography: Chris Menges
Editor: Michael Horton
With Michael Beck, Annie McEnroe,
Bruno Lawrence, John Bach, Diana Rowan,
Kelly Johnson, Ross Jolly, Mark Hadlow,
John Banas, Marshall Napier

Carry Me Back
Director: John Reid
Producer: Graeme Cowley
Screenplay: Derek Morton, Keith Aberdein,
John Reid; from an idea by Joy Cowley
Director of photography: Graeme Cowley
Editors: Simon Reece and Michael Horton
With Grant Tilly, Kelly Johnson, Dorothy McKegg,
Derek Hardwick, Joanne Mildenhall,
Alex Trousdell, Frank Edwards, John Bach,
Fiona Samuel, Peter Tait

Hang On A Minute Mate *(16-millimetre*
telemovie)
Director, producer and screenplay: Alan Lindsay;
based on stories by Barry Crump
Lighting cameraman: Peter Read
Editor: Jamie Selkirk

With Alan Jervis, Kelly Johnson, Alex Trousdale

The Scarecrow
Official selection, Directors Fortnight, Cannes
Film Festival
Director: Sam Pillsbury
Producer: Rob Whitehouse
Screenplay: Michael Heath, Sam Pillsbury; based
on the novel by Ronald Hugh Morrieson
Director of photography: Jim Bartle
Editor: Ian John
With John Carradine, Jonathan Smith,
Tracey Mann, Daniel McLaren, Des Kelly,
Anne Flannery, Jonathan Hardy, Martyn
Sanderson, Greer Robson, Roy Billing,
Sarah Smuts-Kennedy, Yvonne Lawley

Wild Horses
Director: Derek Morton
Producer: John Barnett
Screenplay: Kevin O'Sullivan
Director of photography: Doug Milsome
Editor: Simon Reece
With Keith Aberdein, John Bach, Kevin J. Wilson,
Tom Poata, Bruno Lawrence, Marshall Napier,
Martyn Sanderson

1983

A Woman Of Good Character *(16-millimetre)*
Director: David Blyth
Producer: Grahame McLean
Screenplay: Elizabeth Gowans
Director of photography: John Earnshaw
Editor: Jamie Selkirk
With Sarah Peirse, Jeremy Stephens,
Derek Hardwick, Bruno Lawrence,
Martyn Sanderson, Ian Watkin

Patu! *(documentary)*
Director and producer: Merata Mita
Co-ordinators: Gaylene Preston, Gerd Pohlmann,
Martyn Sanderson

Savage Islands (US title **Nate And Hayes**)
Director: Ferdinand Fairfax
Producers: Rob Whitehouse and Lloyd Phillips
Screenplay: John Hughes, David Odell; based on
a story by Lloyd Phillips
Director of photography: Tony Imi
Editor: John Shirley
With Tommy Lee Jones, Michael O'Keefe, Grant
Tilly, Bill Johnson, Kate Harcourt, Peter Rowley,
David Letch, Bruce Allpress, Reg Ruka, Roy Billing,
Peter Vere-Jones, Mark Hadlow, Phillip Gordon,
Prince Tui Teka

Strata
Director: Geoff Steven
Producer: John Maynard
Executive producer: Gary Hannam
Screenplay: Ester Krumbachova, Geoff Steven,
Michael Havas
Director of photography: Leon Narbey
Editor: David Coulson
With Nigel Davenport, John Banas, Judy Morris,
Roy Billing, Peter Nicoll, Mary Regan,
Phillip Holder, Phillip Gordon, Patrick Smyth

The Lost Tribe
Director and screenplay: John Laing
Producers: Gary Hannam, John Laing
Director of photography: Thomas Burstyn
Editor: Philip McDonald
With John Bach, Darien Takle, Don Selwyn,
Terry Connolly, Ian Watkin, Martyn Sanderson,
Adele Chapman, Joanne Simpson

Utu
Official selection out of competition, Cannes
Film Festival
Director: Geoff Murphy
Producers: Geoff Murphy and Don Blakeney
Screenplay: Geoff Murphy, Keith Aberdein
Director of photography: Graeme Cowley
Editor: Michael Horton

With Anzac Wallace, Bruno Lawrence,
Kelly Johnson, Tim Elliot, John Bach,
Martyn Sanderson, Merata Mita, Tania Bristowe,
Wi Kuki Kaa, Ilona Rogers

War Years (documentary)
Director: Pat McGuire
Producer: Hugh MacDonald
National Film Unit

1984

Among The Cinders
Director: Rolf Haedrich
Producer: John O'Shea
Screenplay: Rolf Haedrich, John O'Shea; based
on the novel by Maurice Shadbolt
Director of photography: Rory O'Shea
Editor: John Kiley
With Paul O'Shea, Derek Hardwick, Yvonne Lawley,
Bridget Armstrong, Rebecca Gibney,
Amanda Jones, Maurice Shadbolt, Des Kelly,
Tom Poata, Peter Baldock, Cherie O'Shea,
Sal Criscillo

Constance
Director: Bruce Morrison
Producer: Larry Parr
Screenplay: Jonathan Hardy and Bruce Morrison
Director of photography: Kevin Hayward
Editor: Melanie Read
With Donogh Rees, Shane Briant, Hester Joyce,
Martin Vaughan, Judie Douglass, Lee Grant,
Donald McDonald, Graham Harvey,
Jonathan Hardy, Yvonne Lawley, Beryl Te Wiata

Death Warmed Up
Director: David Blyth
Producer: Murray Newey
Screenplay: Michael Heath, David Blyth
Director of photography: James Bartle
Editor: David Hugget

With Michael Hurst, Margaret Umbers,
William Upjohn, Norelle Scott, David Letch,
David Weatherley, Jonathan Hardy, Gary Day

Heart Of The Stag
Director: Michael Firth
Producers: Don Reynolds, Michael Firth
Screenplay: Neil Illingworth, with additional
writing by Bruno Lawrence, Michael Firth,
Martyn Sanderson
Director of photography: James Bartle
Editor: Michael Horton
With Bruno Lawrence, Mary Regan,
Terence Cooper, Anne Flannery, Michael Wilson,
Suzanne Cowie, John Bach

Hot Target (aka **Restless**)
Director: Denis Lewiston
Producers: John Barnett, Brian Cook
Director of photography: Alec Mills
Editor: Michael Horton
With Simone Griffith, Steve Marachuk,
Bryan Marshall, Peter McCauley, Elizabeth
Hawthorne, Ray Henwood, Elizabeth McRae,
Vivien Laube, Frank Whitten, Terence Cooper,
Judy McIntosh

Iris (telemovie, videotape)
Director: Tony Isaac
Producers: John Barnett, Tony Isaac
Screenplay: Keith Aberdein
Directors of photography: James Bartle,
Howard Anderson
Editor: Michael Horton
With Helen Morse, Philip Holder, John Bach,
David Aston, Roy Billing, Donogh Rees, Liz McRae

Other Halves
Director: John Laing
Producers: Tom Finlayson, Dean Hill
Screenplay: Sue McCauley, from her novel of the
same name

Director of photography: Leon Narbey
Editor: Harley Oliver
With Lisa Harrow, Mark Pilisi, Paul Gittins,
Temuera Morrison, Alison Routledge, John Bach,
Grant Tilly, Clare Clifford, Bruce Purchase,
Emma Piper

Pallet On The Floor
Director: Lynton Butler
Producer: Larry Parr
Screenplay: Martyn Sanderson, Lynton Butler,
Robert Rising; based on the novel by
Ronald Hugh Morrieson
Director of photography: Kevin Hayward
Editor: Patrick Monaghan
With Peter McCauley, Jillian O'Brien,
Bruce Spence, Shirley Gruar, Alistair Douglas, John
Bach, Marshall Napier, Tony Barry,
Jeremy Stephens, Terence Cooper

The Silent One
Director: Yvonne Mackay
Producer: Dave Gibson
Screenplay: Ian Mune; based on the novel by
Joy Cowley
Director of photography: Ian Paul
Editor: Jamie Selkirk
With Telo Malese, George Henare, Pat Evison,
Anzac Wallace, Reg Ruka, Bernard Kearns,
Prince Tui Teka

Trespasses (working title **Finding Katie**)
Director: Peter Sharp
Producers: Tom Finlayson, Dean Hill
Screenplay: Maurice Gee, Tom Finlayson
Director of photography: Leon Narbey
Editor: David Coulson
With Patrick McGoohan, Emma Piper,
Andy Anderson, Terence Cooper, Frank Whitten,
Sean Duffy, Don Selwyn

Trial Run
Director: Melanie Read
Producer: Don Reynolds
Screenplay: Melanie Read; from an idea by
Caterina de Nave and Melanie Read
Director of photography: Allen Guilford
Editor: Finola Dwyer
With Annie Whittle, Christopher Broun,
Judith Gibson, Stephen Tozer, Martyn Sanderson,
Phillippa Mayne, Lee Grant, Allison Roe,
Maggie Eyre

Vigil
Official selection in competition, Cannes Film
Festival
Director: Vincent Ward
Producer: John Maynard
Executive producer: Gary Hannam
Screenplay: Vincent Ward, Graeme Tetley
Director of photography: Alun Bollinger
Editor: Simon Reece
With Fiona Kay, Bill Kerr, Penelope Stewart,
Frank Whitten, Gordon Shields

1985

Came A Hot Friday
Director: Ian Mune
Producer: Larry Parr
Screenplay: Dean Parker, Ian Mune; from the
novel by Ronald Hugh Morrieson
Director of photography: Alun Bollinger
Editor: Ken Zemke
With Peter Bland, Phillip Gordon,
Michael Lawrence, Billy T. James, Marshall Napier,
Marise Wipani, Erna Larsen, Don Selwyn,
Phillip Holder, Bruce Allpress, Tricia Phillips,
Michael Morrisey, Bridget Armstrong

Kingpin
Director: Mike Walker
Producers: Gary Hannam, Mike Walker

Screenplay: Mike Walker, Mitchell Manuel
Director of photography: John Toon
Editor: Paul Sutorius
With Mitchell Manuel, Junior Amiga,
Judy McIntosh, Jim Moriarty, Terence Cooper,
Wi Kuki Kaa, Peter McCauley, Nicholas Rogers

Leave All Fair *(filmed in France)*
Director: John Reid
Producer: John O'Shea
Screenplay: Stanley Harper, Maurice Pons,
Jean Betts, John Reid
Director of photography: Bernard Lutic
Editor: Ian John
With Jane Birkin, John Gielgud, Feodor Atkine,
Simon Ward

Mr Wrong
Director: Gaylene Preston
Producers: Robin Laing and Gaylene Preston
Associate producer: Don Reynolds
Screenplay: Gaylene Preston with Geoff Murphy
and Graeme Tetley; based on the story by
Elizabeth Jane Howard
Director of photography: Thom Burstyn
Camera operator: Alun Bollinger
Editor: Simon Reece
With Heather Bolton, David Letch,
Margaret Umbers, Suzanne Lee, Danny Mulheron,
Gary Stalker, Perry Piercy, Philip Gordon,
Michael Haigh, Kate Harcourt

Queen City Rocker
Director: Bruce Morrison
Producer: Larry Parr
Screenplay: Bill Baer
Director of photography: Kevin Hayward
Editor: Michael Hacking
With Matthew Hunter, Mark Pilisi,
Kim Willoughby, Peter Bland, George Henare,
Greer Robson, Rebecca Saunders, Liddy Holloway,
Joel Tobeck

Shaker Run

Director: Bruce Morrison

Producers: Larry Parr, Igo Cantor

Screenplay: James Kouf Jnr, Henry Fownes, Bruce Morrison

Director of photography: Kevin Hayward

Editors: Ken Zemke, Bob Richardson

With Cliff Robertson, Lisa Harrow, Leif Garrett, Shane Briant, Peter Rowell, Peter Hayden, Ian Mune, Bruce Phillips, Fiona Samuels, Nathaniel Lees

Should I Be Good?

Producer, director and screenplay: Grahame McLean

Director of photography: Warrick Attewell

Editor: Jamie Selkirk

With Hammond Gamble, Harry Lyon, Joanne Mildenhall, Beaver, Spring Rees, Terence Cooper

Sylvia

Director: Michael Firth

Producers: Don Reynolds, Michael Firth

Screenplay: Michele Quill, F. Fairfax, Michael Firth; based on *Teacher* and *I Passed This Way* by Sylvia Ashton-Warner

Director of photography: Ian Paul

Editor: Michael Horton

With Eleanor David, Nigel Terry, Tom Wilkinson, Mary Regan, Martyn Sanderson, Terence Cooper, Sarah Peirse, David Letch

The Lie Of The Land

Director: Grahame McLean

Producers: Grahame McLean, Narelle Barsby

Writer: Grahame McLean; based on an original screenplay *Small Farms* by Kevin Smith

Director of photography: Warrick Attewell

Editor: Jamie Selkirk

With Marshall Napier, Jim Moriarty, Robert Wallach, Terence Cooper, Jonathan Hardy, Dean Moriarty, Ann Pacey, Tom Poata, John Bach

The Neglected Miracle (documentary)

Director: Barry Barclay

Producers: John O'Shea, Craig Walters

Director of photography: Rory O'Shea

Editor: Annie Collins

The Quiet Earth

Director: Geoff Murphy

Producers: Sam Pillsbury and Don Reynolds

Screenplay: Sam Pillsbury, Bill Baer, Bruno Lawrence; based on the novel by Craig Harrison

Director of photography: James Bartle

Editor: David Hugget

With Bruno Lawrence, Alison Routledge, Peter Smith, Anzac Wallace

1986

Arriving Tuesday

Director: Richard Riddiford

Producers: Don Reynolds, Chris Hampson

Screenplay: Richard Riddiford, David Copeland

Director of photography: Murray Milne

Editor: John McWilliam

With Judy McIntosh, Rawiri Paratene, Peter Hayden, Heather Bolton, Sarah Peirse, Lee Grant, Frank Whitten

Bridge To Nowhere

Director: Ian Mune

Producer: Larry Parr

Screenplay: Ian Mune, Bill Baer

Director of photography: Kevin Hayward

Editor: Finola Dwyer

With Bruno Lawrence, Margaret Umbers, Alison Routledge, Phillip Gordon, Matthew Hunter, Shelley Luxford, Stephen Judd

Dangerous Orphans
Director: John Laing
Producer: Don Reynolds
Screenplay: Kevin Smith
Director of photography: Warrick Attewell
Editor: Michael Horton
With Michael Hurst, Peter Stevens, Ross Girven,
Jennifer Ward-Lealand, Peter Bland, Grant Tilly,
Zac Wallace, Ian Mune, Ann Pacey,
Peter Vere-Jones

Footrot Flats: The Dog's (Tail) Tale
Director: Murray Ball
Producers: John Barnett, Pat Cox
Editors: Michael Horton, Denis Jones
Screenplay: Murray Ball, Tom Scott; based on
characters created by Murray Ball

The Fire-Raiser *(telemovie)*
Director: Peter Sharp
Producer: Ginette McDonald

Worzel Gummidge Down Under *(telemovie)*
Director: James Hill
Producer: Grahame McLean
Screenplay: Keith Waterhouse and Willis Hall
With John Pertwee, Una Stubbs, Jonathan Marks,
Olivia Ihimaera Smiler

1987

Mark II *(telemovie, video)*
Director: John Anderson
Producer: Dan McKirdy
Screenplay: Mike Walker, Mitchell Manuel
With Mitchell Manuel, Junior Amiga,
Nicholas Rogers, Jim Moriarty, Riwia Brown,
Tama Poata, Joanna Briant, Jeff Boyd,
Kate Harcourt

Ngati
Official selection, Critics Week, Cannes
Film Festival

Director: Barry Barclay
Producer: John O'Shea
Associate producers: Craig Walters, Tama Poata
Screenplay: Tama Poata
Director of photography: Rory O'Shea
Editor: Dell King
With Wi Kuki Kaa, Tuta Ngarimu Tamati,
Connie Pewhairangi, Michael Tibble, Oliver Jones,
Ross Girven, Judy McIntosh, Kiri McCorkindale,
Norman Fletcher, Lucky Renata, Paki Cherrington,
Ngawai Harris, Tawai Moana, Alice Fraser

Starlight Hotel
Director: Sam Pillsbury
Producers: Larry Parr, Finola Dwyer
Screenplay: Grant Hindin-Miller from his novel
The Dream Mongers
Director of photography: Warrick Attewell
Editor: Michael Horton
With Greer Robson, Peter Phelps, Marshall Napier,
Alice Fraser, Patrick Smyth, Donogh Rees,
Gary McCormick, The Wizard, Bruce Phillips

The Leading Edge
Director: Michael Firth
Producer: Barrie Everard
Screenplay: Michael Firth, Grant Morris
Directors of photography: Stuart Dryburgh,
Michael Firth
Editor: Patrick Monaghan
With Mathurin Molgat, Bruce Grant,
Evan Bloomfield, Christine Grant, Mark Whetu,
Melanie Forbes, Billy T. James

1988

A Soldier's Tale *(filmed in France)*
Director-producer: Larry Parr
Executive producer: Don Reynolds
Screenplay: Larry Parr, Grant Hindin-Miller;
based on the novel by M.K. Joseph
Director of photography: Alun Bollinger

Editor: Michael Horton
With Gabriel Byrne, Marianne Basler, Judge
Reinhold, Paul Wyett, Benoit Regent

Bad Taste
Director, producer, screenplay and camera:
Peter Jackson
Editors: Peter Jackson, Jamie Selkirk
With Peter Jackson, Pete O'Herne, Terry Potter,
Craig Smith, Mike Minett, Doug Wren

Illustrious Energy
Director: Leon Narbey
Producers: Don Reynolds, Chris Hampson
Screenplay: Leon Narbey, Martin Edmond
Director of photography: Alan Locke
Editor: David Coulson
With Shaun Bao, Harry Ip, Peter Chin, Geeling,
Desmond Kelly, Heather Bolton, Peter Hayden,
David Telford

In Our Own Time (documentary)
Director and producer: Andrea Bosshard,
Shane Loader, Jeremy Royal

Mauri
Director, producer and screenplay: Merata Mita
Associate producer: Geoff Murphy
Director of photography: Graeme Cowley
Editor: Nicholas Beauman
With Eva Rickard, Anzac Wallace, Geoff Murphy,
Don Selwyn, Temuera Morrison, James Heyward,
Susan D. Ramiri Paul, Rangimarie Delamare,
Sonny Waru, Willie Raana

Never Say Die
Director and screenplay: Geoff Murphy
Producers: Geoff Murphy, Murray Newey
Executive producer: Barrie Everard
Director of photography: Rory O'Shea
Editor: Scott Conrad
With Temuera Morrison, Lisa Eilbacher, Tony Barry,
George Wendt, Geoff Murphy, Barrie Everard,

Murray Newey, Alan Sorrell, Phillip Gordon,
Martyn Sanderson, Sean Duffy, Tom Poata,
Gay Dean

Send A Gorilla
Director and screenplay: Melanie Read; from an
idea by Perry Piercy
Producer: Dorothee Pinfold
Director of photography: Wayne Vinten
Editor: Paul Sutorius
With Carmel McGlone, Katherine McRae,
Perry Piercy, John Callen, Larney Tupu,
Jim Moriarty, Rima Te Wiata

The Grasscutter (telemovie)
Director: Ian Mune
Producer: Tom Finlayson
Screenplay: Roy Mitchell
Directors of photography: Matt Bowkett,
Michael O'Connor
Editor: Patrick Monaghan
With Mitchell Manuel, Judy McIntosh,
Terence Cooper, Martin Maguire, Ian McElhinney,
Frances Barber, Marshall Napier, Jon Brazier

The Navigator
Official selection in competition, Cannes Film
Festival
Director: Vincent Ward
Producer: John Maynard
Executive producer: Gary Hannam
Screenplay: Vincent Ward, Kely Lyons,
Geoff Chapple; from an original idea by
Vincent Ward
Director of photography: Geoffrey Simpson
Editor: John Scott
With Bruce Lyons, Chris Hayward,
Hamish McFarlane, Marshall Napier,
Noel Appleby, Paul Livingstone, Sarah Peirse,
Jay Lavea Laga'aia
An official New Zealand-Australia
co-production

The Rainbow Warrior Conspiracy *(telemovie)*
Director: Chris Thomson
Producer: Robert Loader
Screenplay: David Phillips
With Brad Davis, Jack Thompson, Louise Lapare,
Germaine Houde, Mary Regan, Peter Carroll,
Bruno Lawrence, Temuera Morrison, Suzanne Lee,
Andrea Cunningham, Lorae Parry,
Sarah Smuts-Kennedy

1989

Meet The Feebles
Director: Peter Jackson
Producers: Jim Booth, Peter Jackson
Screenplay: Peter Jackson, Fran Walsh,
Stephen Sinclair, Danny Mulheron
Director of photography: Murray Milne
Editor: Jamie Selkirk

The Champion *(telemovie)*
Director: Peter Sharp
Producer: Ginette McDonald
With Don Selwyn, Peter Tait, Milan Borich,
Alistair Douglas

Zilch!
Director: Richard Riddiford
Producers: Amanda Hocquard, Richard Riddiford
Screenplay: Richard Riddiford,
Jonathan Dowling
Director of photography: Murray Milne
Editor: Chris Todd
With Michael Mizrahi, Lucy Sheehan,
John Watson, Roy Billing, Peter Tait,
Edward Campbell, Andy Anderson, Frank Whitten

1990

An Angel At My Table
Official selection in competition, Venice Film
Festival: winner Special Jury Award
Director: Jane Campion

Producer: Bridget Ikin
Co-producer: John Maynard
Screenplay: Laura Jones; based on the
autobiographies of Janet Frame
Director of photography: Stuart Dryburgh
Editor: Veronika Haussler
With Kerry Fox, Alexia Keogh, Karen Fergusson,
Iris Churn, Kevin J. Wilson, Melina Bernecker,
Andrew Binns, Glynis Angell,
Sarah Smuts-Kennedy, Martyn Sanderson,
David Letch, William Brandt, Edith Campion,
Brenda Kendall

Flying Fox In A Freedom Tree
Director and screenplay: Martyn Sanderson;
based on the novella by Albert Wendt
Producer: Grahame McLean
Director of photography: Allen Guilford
Editor: Ken Zemke
With Faifua Amiga Jnr, Richard von Sturmer

Mana Waka *(documentary)*
Director: Merata Mita
Photography (1937–1940): R.G.H. Manley
Narrator: Tukuroirangi Morgan

Ruby And Rata
Director: Gaylene Preston
Producers: Robin Laing, Gaylene Preston
Screenplay: Graeme Tetley
Director of photography: Leon Narbey
Editor: Paul Sutorius
With Yvonne Lawley, Vanessa Rare, Simon Barnett,
Lee Mete-Kingi, Debes Bhattacharyya,
Russell Gowers, Russell Smith, Ngaire Horton,
Alma Woods, Vicky Burrett, Steve Lahood

The Rogue Stallion *(aka **Wildfire**) (telemovie)*
Director: Henri Safran
Producers: Don Reynolds, Philip East
Executive producer: Roger Mirams

With Bruno Lawrence, Rawiri Paratene, Beaver,
Jodie Rimmer, Dean O'Gorman
An official New Zealand-Australia
co-production

User Friendly
Director: Gregor Nicholas
Producers: Trevor Haysom, Frank Stark
Screenplay: Gregor Nicholas, Norelle Scott,
Frank Stark
Director of photography: Donald Duncan
Editor: David Coulson
With William Brandt, Alison Bruce, Judith Gibson,
David Letch, Lewis Martin, Joan Reid, June Bishop,
Noel Appleby, Neil Weatherley, Dorothy Hurt,
Belinda Weymouth

1991

Chunuk Bair
Director: Dale Bradley
Producer: Grant Bradley
Screenplay: Grant Hindin-Miller; based on the
play *Once on Chunuk Bair* by Maurice Shadbolt
Director of photography: Warrick Attewell
Editor: Paul Sutorius
With Robert Powell, Kevin J. Wilson, Jed Brophy,
Richard Hanna, Peter Kaa, John Leigh,
Murray Keane, Stephen Ure, Darryl Beattie,
Tim Bray

Old Scores
Director: Alan Clayton
Producer: Don Reynolds
Screenplay: Dean Parker, Greg McGee
Director of photography: Allen Guilford
Editors: Michael Horton, Jamie Selkirk
With John Bach, Tony Barry, Roy Billing,
Alison Bruce, Stephen Tozer, Robert Bruce,
Terence Cooper, Martyn Sanderson, Waka Nathan,
Keith Quinn, Windsor Davies

Rebels In Retrospect *(documentary, beta)*
Director and producer: Russell Campbell

Te Rua
Director and screenplay: Barry Barclay
Producer: John O'Shea
Directors of photography: Rory O'Shea,
Warrick Attewell
Editors: Simon Reece, Dell King
With Wi Kuki Kaa, Matiu Mareikura, Peter Kaa,
Dalvanius, Donna Akersten, Nissie Herewini,
Tilly Reedy, Gunter Meisner, Maria Fitzi,
Stuart Devenie

The End Of The Golden Weather
Director: Ian Mune
Producers: Christina Milligan, Ian Mune
Executive producer: Don Reynolds
Screenplay: Ian Mune and Bruce Mason; from
the stage play by Bruce Mason
Director of photography: Alun Bollinger
Editor: Michael Horton
With Stephen Fulford, Stephen Papps, Paul Gittins,
Gabrielle Hammond, Ray Henwood, Alice Fraser,
Bill Johnson, Steve McDowell, Alistair Douglas,
Greg Johnson

The Returning
Director: John Day
Producer: Trishia Downie
Screenplay: Arthur Baysting, John Day
Director of photography: Kevin Hayward
Editor: Simon Clothier
With Phillip Gordon, Alison Routledge,
Tony Groser, Jim Moriarty, Shirley Grace,
Grant Tilly, Max Cullen, John Ewart,
Frank Whitten, Judie Douglass

Undercover *(telemovie, video)*
Director: Yvonne Mackay
Producer: Dave Gibson
Screenplay: Arthur Baysting

Director of photography: Wayne Vinten
Editor: Mike Bennett
With William Brandt, Jennifer Ludlum,
Hori Ahipene, Alistair Browning, Gerard Tahu,
Cliff Curtis, James Heyward

1992

Alex
Director: Megan Simpson
Producers: Tom Parkinson (NZ), Phil Gerlach
(Australia)
Screenplay: Ken Catran, adapted from the book
by Tessa Duder
Director of photography: Donald Duncan
Editor: Tony Kavanagh
With Lauren Jackson, Chris Haywood, Josh Picker,
Elizabeth Hawthorne, Bruce Phillips,
Catherine Godbold, Grant Tilly
An official New Zealand-Australia co-production

Braindead
Director: Peter Jackson
Producer: Jim Booth
Associate producer: Jamie Selkirk
Screenplay: Peter Jackson, Fran Walsh,
Stephen Sinclair; from an original idea by
Stephen Sinclair
Director of photography: Murray Milne
Editor: Jamie Selkirk
With Timothy Balme, Diana Penalver,
Elizabeth Moody, Stuart Devenie, Ian Watkin,
Brenda Kendall, Jed Brophy, Davina Whitehouse,
Murray Keane

Crush
Official selection in competition, Cannes Film
Festival
Director: Alison Maclean
Producer: Bridget Ikin
Associate producer: Trevor Haysom
Screenplay: Alison Maclean, Anne Kennedy
Director of photography: Dion Beebe

Editor: John Gilbert
With Marcia Gay Harden, Donogh Rees,
Caitlin Bossley, William Zappa, Jon Brazier

Marlin Bay *(telemovie)*
Director: Chris Bailey
Producer: Janice Finn
Screenplay: Greg McGee
With Ilona Rogers, Andy Anderson, Lani John
Tupu, May Lloyd, Katie Woolf, Rupert Green,
Don Selwyn, Kenneth Blackburn, Gilbert Goldie

Moonrise *(aka* Grampire*)* *(US title* My Grandpa Is A Vampire*)*
Director: David Blyth
Producer: Murray Newey
Associate producers: Judith Trye, Bryan Walden
Screenplay: Michael Heath; from his radio play
of the same name
Director of photography: Kevin Hayward
Editor: David Huggett
With Al Lewis, Justin Gocke, Milan Borich,
Pat Evison, Noel Appleby, David Weatherly,
Sean Duffy, Sylvia Rand, Beryl Te Wiata,
Alistair Douglas, Max Cryer

The Footstep Man
Director: Leon Narbey
Producer: John Maynard
Associate producer: Trevor Haysom
Screenplay: Leon Narbey and Martin Edmond
Director of photography: Allen Guilford
Editor: David Coulson
With Michael Hurst, Jennifer Ward-Lealand,
Steven Grives, Sarah Smuts-Kennedy, Rosey Jones,
Harry Sinclair, Glenis Levestam, Jessica Maynard

1993

Absent Without Leave
Director: John Laing
Producer: Robin Laing
Screenplay: Graeme Tetley, Jim Edwards

Director of photography: Allen Guilford
Editor: Paul Sutorius
With Craig McLachlan, Katrina Hobbs, Tony Barry,
Judie Douglass, Robyn Malcolm, David Copeland,
Ken Blackburn, Desmond Kelly, Chloe Laing

Bread And Roses *(16-millimetre)*
Director: Gaylene Preston
Producer: Robin Laing
Executive producer: Dorothee Pinfold
Screenplay: Gaylene Preston and Graeme Tetley;
from *Bread and Roses* by Sonja Davies
Director of photography: Allen Guilford
Editor: Paul Sutorius
With Genevieve Picot, Donna Akersten, Mick Rose,
Raymond Hawthorne, Tina Retgien, Erik Thomson,
Theresa Healey

Cops And Robbers
Director: Murray Reece
Producers: Tom Parkinson (NZ), Phil Gerlach
(Australia)
Screenplay: Timothy Bean
Director of photography: Steve Arnold
Editor: Simon Reece
With Rima Te Wiata, Mark Wright
An official New Zealand-Australia
co-production

Desperate Remedies
Official selection, Un Certain Regard, Cannes
Film Festival
Directors and writers: Peter Wells and Stewart
Main
Producer: James Wallace
Associate producer: Trishia Downie
Director of photography: Leon Narbey
Editor: David Coulson
With Jennifer Ward-Lealand, Kevin Smith,
Lisa Chappell, Cliff Curtis, Michael Hurst,
Bridget Armstrong, Kiri Mills

Jack Be Nimble
Director and screenplay: Garth Maxwell
Producers: Jonathan Dowling, Kelly Rogers
Associate producer: Judith Trye
Director of photography: Donald Duncan
Editor: John Gilbert
With Alexis Arquette, Sarah Smuts-Kennedy,
Elizabeth Hawthorne, Tony Barry, Bruno Lawrence

The Piano
Official selection in competition, Cannes Film
Festival: co-winner, Palme d'Or
Academy Awards: best original screenplay, best
actress, best supporting actress
Director and screenplay: Jane Campion
Producer: Jan Chapman
Director of photography: Stuart Dryburgh
Editor: Veronika Jenet
With Sam Neill, Holly Hunter, Anna Paquin,
Harvey Keitel, Kerry Walker, Genevieve Lemon,
Tungia Baker, Ian Mune, Pete Smith,
Bruce Allpress, Cliff Curtis, Hori Ahipene

1994

Heavenly Creatures
Official selection in competition, Venice Film
Festival: winner Silver Lion
Academy Award nomination, best original
screenplay
Director: Peter Jackson
Producer: Jim Booth
Executive producer: Hanno Huth
Screenplay: Peter Jackson, Fran Walsh
Director of photography: Alun Bollinger
Editor: Jamie Selkirk
With Melanie Lynskey, Kate Winslet, Diana Kent,
Clive Merrison, Sarah Peirse, Jed Brophy,
Elizabeth Moody, Simon O'Connor

Loaded *(filmed in England)*
Director and screenplay: Anna Campion
Producers: John Maynard and Bridget Ikin (NZ),
David Hazlett, Caroline Hewitt (UK)
Director of photography: Alan Almond
Editor: John Gilbert
An official New Zealand-United Kingdom
co-production

Once Were Warriors
Official selection, Venice Film Festival: best
first film
Director: Lee Tamahori
Producer: Robin Scholes
Screenplay: Riwia Brown; based on the novel
by Alan Duff
Director of photography: Stuart Dryburgh
Editor: Michael Horton
*With Temuera Morrison, Rena Owen,
Mamaengaroa Kerr-Bell, Julian Arahanga,
Taungaroa Emile, Rachael Morris Jnr, Joseph
Kairua, Cliff Curtis, Pete Smith, George Henare,
Mere Boynton, Shannon Williams*

The Last Tattoo
Director: John Reid
Producers: Neville Carson and Bill Gavin
Screenplay: Keith Aberdein; from a story
by John Reid and Keith Aberdein
Director of photography: John Blick
Editor: John Scott
*With Kerry Fox, Tony Goldwyn, Robert Loggia,
Rod Steiger, John Bach, Timothy Balme, Tony Barry,
Elizabeth Hawthorne, Peter Hambleton,
Martyn Sanderson, Katie Wolfe, Desmond Kelly,
Donna Akersten, Eddie Campbell,
Danielle Cormack, Michelle Huirama*

Vulcan Lane *(video)*
Producer-director: Michael Firth

1995

Bonjour Timothy
Director: Wayne Tourell
Producers: Murray Newey, Judith Trye (NZ),
Micheline Charest, Patricia Levoie (Canada)
Screenplay: David Preston
Director of photography: Matt Bowkett
Editor: Jean-Marie Drot
*With Dean O'Gorman, Sabine Karsenti,
Angela Bloomfield, Syd Jackson, Sylvia Rands,
Richard Vette, Mark Hadlow, Stephen Papps*
An official New Zealand-Canada
co-production

Flight Of The Albatross
Director: Werner Meyer
Producers: Vincent Burke (NZ), Udo Heiland
(Germany)
Screenplay: Riwia Brown; based on the book
by Deborah Savage
Director of photography: Martin Gressman
Editor: Michael Horton
*With Taungaroa Emile, Diana Ngaromotu-Heka,
Julia Brendler, Jack Thompson, Pete Smith,
Peter Tait, Joan Reid, Eva Rickard, Grant Tilly,
Beryl Te Wiata*

Jack Brown, Genius
Director: Tony Hiles
Producer: Jamie Selkirk
Executive producers: Peter Jackson, Hanno Huth
Development producer: Jim Booth
Associate producer: Sue Rogers
Screenplay: Tony Hiles, Peter Jackson, Fran Walsh;
from an original idea by Tony Hiles
Director of photography: Allen Guilford
Editor: Jamie Selkirk
*With Timothy Balme, Stuart Devenie,
Lisa Chappell, Nicola Murphy, Marton Csokas,
Edward Campbell*

War Stories (War Stories Our Mothers Never Told Us) (documentary)
Official selection, Venice Film Festival
Director and producer: Gaylene Preston
Executive producer: Robin Laing
Director of photography: Alun Bollinger
Editor: Paul Sutorius
With Tui Preston, Pamela Quill, Neva Clarke McKenna, Flo Small, Mabel Waititi, Rita Graham, Jean Andrews; interviewer: Judith Fyfe

1996

Broken English
Director: Gregor Nicholas
Producer: Robin Scholes
Executive producer: Timothy White
Screenplay: Gregor Nicholas, Johanna Piggott, Jim Salter
Director of photography: John Toon
Editor: David Coulson
With Julian Arahanga, Marton Csokas, Temuera Morrison, Rade Serbedzija, Aleksandra Vujcic, Madeline McNamara, Elizabeth Mavric

Chicken
Director and screenplay: Grant Lahood
Producer: John Keir
Executive producer: Hanno Huth
Director of photography: Allen Guilford
Editor: John Gilbert
With Bryan Marshall, Ellie Smith, Martyn Sanderson, Cliff Curtis, Jed Brophy, Claire Waldron, Joan Dawe, Sean Feehan

Someone Else's Country (documentary, beta-sp)
Director and researcher: Alister Barry
Photography: Andrea Bosshard, Jonathan Brough, Mark Derby, Shane Loader, Peter DaVanzo, Tony Sutorius, Barry Thomas
Editor: Shane Loader
Narrator: Ian Johnstone

The Frighteners
Director: Peter Jackson
Producers: Peter Jackson, Jamie Selkirk
Executive producer: Robert Zemeckis
Screenplay: Fran Walsh, Peter Jackson
Directors of photography: Alun Bollinger, John Blick
Editor: Jamie Selkirk
With Michael J. Fox, Trini Alvorado, Elizabeth Hawthorne, Jonathan Blick, Peter Dobson, Angela Bloomfield, Stuart Devenie, Desmond Kelly, Ken Blackburn, Melanie Lynskey

The Whole Of The Moon
Director: Ian Mune
Producers: Murray Newey (NZ), Micheline Charest (Canada)
Screenplay: Ian Mune, Richard Lymposs
Director of photography: Warrick Attewell
Editor: Jean Beaudoin
With Toby Fisher, Nikki Si'Ulepa, Pascal Bussieres, Carl Bland
An official New Zealand-Canada co-production

1997

Aberration
Director: Tim Boxell
Producers: Chris Brown, Tim Sanders
Director of photography: Allen Guilford
Editor: John Gilbert

Return To Treasure Island (telemovie)
Director: Steve Lahood

Saving Grace
Director: Costa Botes
Producer: Larry Parr
Screenplay: Duncan Sarkies, based on his play
Director of photography: Sean O'Donnell
Editor: Michael Horton
With Jim Moriarty, Kirsty Hamilton

The Climb

Director: Bob Swaim
NZ producer: Tom Parkinson
Director of photography: Allen Guilford
With John Hurt, Gregory Smith, David Strathairn,
Tina Regtien, Michael Galvin, Peter Rowley
An official New Zealand-Canada-France
co-production

The Road To Jerusalem *(documentary, beta-sp)*

Director: Bruce Morrison
Producer: William Grieve
Screenplay: Paul Millar, Bruce Morrison;
featuring the words of James K. Baxter
With Jacqui Baxter, John Baxter, Millicent Baxter,
Terence Baxter, Brian Bell, Colin Durning,
Noel Ginn, Sam Hunt, Mike Minehan

The Ugly

Director and screenplay: Scott Reynolds
Producer: Jonathan Dowling
Director of photography: Simon Raby
Editor: Wayne Cook
With Paolo Rotondo, Rebecca Hobbs, Roy Ward,
Jennifer Ward-Lealand, Darien Takle, Paul Glover

Topless Women Talk About Their Lives

Director and screenplay: Harry Sinclair
Producer: Fiona Copland
Director of photography: Dale McCready
Editor: Cushla Dillon
With Danielle Cormack, Ian Hughes, Joel Tobeck,
Willa O'Neill, Shimpal Lelisi, Andrew Binns,
Peter Elliott, Tania Simon, Ian Mune,
Frances Edmond

1998

Heaven

Director and screenplay: Scott Reynolds;
based on the novel by Chad Taylor
Producer: Sue Rogers
Director of photography: Simon Raby
Editor: Wayne Cook
With Martin Donovan, Richard Schiff, Karl Urban

Memory And Desire

Official selection, Critics Week, Cannes Film
Festival
Director and screenplay: Niki Caro; based on a
story by Peter Wells
Producer: Owen Hughes
Director of photography: Dion Beebe
Editor: Margot Francis
With Yuri Kinugawa, Eugene Nomura, Joel Tobeck

The Lost Valley *(aka **Kiwi Safari**)*

Director, screenplay: Dale Bradley
Producer: Grant Bradley
Photography: Neil Cervin
Editors: Douglas Braddock, Dale Bradley
With Meg Foster, Andrea Thompson,
Kevin J. Wilson, Stephen Hall, Tamati Rice

The Lunatics' Ball

Director, screenplay, producer: Michael Thorp
Photography: Neil Cervin
Editors: Michael Horton, Paul Sutorius
With Russel Walder, Jane Irwin, Sarah Ashworth,
Michael Daley, Alan De Malmanche, Jon Brazier

Tiger Country *(telemovie)*

Director: John Laing
Producers: Dave Gibson, Tom Scott

Via Satellite

Director: Anthony McCarten
Producer: Philippa Campbell
Screenplay: Anthony McCarten;
co-writer: Greg McGee
Director of photography: Simon Riera
Editor: John Gilbert
With Danielle Cormack, Rima Te Wiata,
Karl Urban, Tim Balme, Jodie Dorday,
Brian Sergent, Donna Akersten

When Love Comes
Director: Garth Maxwell
Producer: Michele Fantl
Screenplay: Garth Maxwell, Rex Pilgrim,
Peter Wells
Director of photography: Darryl Ward
Editor: Cushla Dillon
With Rena Owen, Dean O'Gorman, Simon Prast,
Sophia Hawthorne, Nancy Brunning,
Simon Westaway

1999

Campaign (documentary, beta-sp)
Director, producer and photography:
Tony Sutorius
Editor: Andrew Mortimer
With Pauline Gardiner, Danna Glendinning,
Richard Prebble, Alick Shaw

Channelling Baby
Director and screenplay: Christine Parker
Producer: Caterina de Nave
Director of photography: Rewa Harre
Editor: Chris Plummer
With Danielle Cormack, Kevin Smith,
Amber Sainsbury, Joel Tobeck

Getting To Our Place (documentary, beta-sp)
Directors: Gaylene Preston, Anna Cottrell
Producer: Gaylene Preston
Photography: Chris Terpstra, Darryn Smith
Editor: Paul Sutorius
With Cheryl Sotheran, Cliff Whiting,
Sir Ron Trotter, Paul Thompson, Jock Phillips

I'll Make You Happy
Director and screenplay: Athina Tsoulis
Producer: Liz Stevens
Screenplay: Athina and Anne Tsoulis
Director of photography: Rewa Harre
Editor: Chris Plummer

With Jodie Rimmer, Sandy Ireland, Carl Bland, Ian
Hughes, Michael Hurst, Jennifer Ward-Lealand,
Rena Owen, Raybon Kan

Punitive Damage (documentary)
Director and script: Annie Goldson
Producers: Annie Goldson and Gaylene Preston
Director of photography: Leon Narbey
Editor: John Gilbert
With Helen Todd, Constancio Pinto, Alan Nairn,
Michael Ratner, Beth Stephens

Savage Honeymoon
Director and screenplay: Mark Beesley
Producer: Steve Sachs
Director of photography: Leon Narbey
Editor: Margot Francis
With Nicholas Eadie, Perry Piercy,
Elizabeth Hawthorne, Craig Hall,
Sophia Hawthorne, Ian Mune

Scarfies
Director: Robert Sarkies
Producer: Lisa Chatfield
Screenplay: Duncan Sarkies, Robert Sarkies
Director of photography: Stephen Downes
Editor: Annie Collins
With Taika Cohen, Willa O'Neill, Ashleigh Segar,
Charlie Bleakley, Jon Brazier, Neill Rea,
Mark Neilson

Uncomfortable, Comfortable (beta-sp)
Director: Campbell Walker
Producers: Campbell Walker, Diane McAllen
Screenplay: Colin Hodson, Robyn Venables,
Tracy Arnold, James Knuckey, Campbell Walker
Photography: Roland Ebbing
Editors: Campbell Walker, Colin Hodson,
Diane McAllen
With Colin Hodson, Robyn Venables, Tracy Arnold,
James Knuckey

What Becomes Of The Broken Hearted?
Director: Ian Mune
Producer: Bill Gavin
Executive producers: John Barnett, Richard
Sheffield
Screenplay: Alan Duff; based on his novel of
the same name
Director of photography: Allen Guilford
Editor: Michael Horton
With Temuera Morrison, Nancy Brunning,
Julian Arahanga, Clint Eruera, Rena Owen,
Tony Arahanga, Lawrence Makaore, Roi Taimana,
Sonny Kirikiri

Wild Blue
Director and screenplay: Dale Bradley
Producer: Grant Bradley
Director of photography: Neil Cervin
Editor: Doug Braddock
With Nicola Murphy, Judge Reinhold,
Morgan Palmer Hubbard

2000

Fearless *(telemovie)*
Director: Charlie Haskell
NZ producer: John Barnett
Executive producers include Roger Donaldson
With Dean O'Gorman and Kelson Henderson
An official New Zealand-Canada co-production

Hopeless
Director: Stephen Hickey
Producer: Larry Parr
Screenplay: Stephen Hickey, Sean Molloy
Director of photography: Simon Reira
Editors: Andrew Brettell,
Jonno Woodford-Robinson
With Phil Pinner, Adam Gardiner,
Mia Taumoepeau, Scott Wills, Sally Stockwell

Jubilee
Director: Michael Hurst
Producer: Bill Gavin
Executive producers: John Barnett,
Caterina de Nave
Screenplay: Michael Bennett; based on the book
by Nepi Solomon
Director of photography: Leon Narbey
Editor: Eric de Beus
With Cliff Curtis, Theresa Healey, Hori Ahipene,
Kevin Smith, Taungaroa Emile, Vicky Haughton,
Elizabeth Hawthorne

Magik And Rose
Director and screenplay: Vanessa Alexander
Producer: Larry Parr
Director of photography: Fred Renata
Editor: Eric de Beus
With Alison Bruce, Nicola Murphy, Oliver Driver,
Simon Ferry, Florence Hartigan

Shifter *(beta-sp)*
Director/editor: Colin Hodson
Producers: Diane McAllen, Campbell Walker,
Colin Hodson
Photography: Campbell Walker
With Colin Hodson, Samara McDowell,
Diane McAllen, Johanna Sanders,
Campbell Walker, Siobhan Garrett

Stickmen
Director: Hamish Rothwell
Producer: Michelle Turner
Screenplay: Nick Ward
Director of photography: Nigel Bluck
Editor: Owen Ferrier-Kerr
With Robbie Magasiva, Scott Wills, Paolo Rotondo,
Simone Kessell, Kirk Torrance, Mick Rose

Street Legal: Hit And Run *(telemovie)*
Director: Chris Bailey
Producer: Chris Hampson

With Owen Black, Jay Laga'aia,
Katherine Kennard, Charles Mesure

The Feathers Of Peace

Director and screenplay: Barry Barclay
Producer: Ruth Kaupua-Panapa
Executive producer: Don Selwyn
Director of photography: Michael O'Connor
Editor: Bella Erikson
With Calvin Tuteao, John Callen, Sonny Kirikiri,
Michael Lawrence, Lawrence Makoare,
Graeme Moran, Eryn Wilson, Michael Holt,
Mick Innes, Roimata Taimana, Lionel Waaka,
Peter Tait, John Paekau, Max Auld

The Irrefutable Truth About Demons

Director and screenplay: Glenn Standring
Producer: Dave Gibson
Director of photography: Simon Baumfield
Editor: Paul Sutorius
With Karl Urban, Sally Stockwell, Katie Wolfe,
Jonathan Hendry, Tony MacIver

The Price Of Milk

Director and screenplay: Harry Sinclair
Producer: Fiona Copland
Executive producer: Tim Sanders
Director of photography: Leon Narbey
Editor: Cushla Dillon
With Danielle Cormack, Karl Urban, Willa O'Neill,
Michael Lawrence, Rangi Motu,
Lawrence Makoare

The Shirt *(beta-sp)*

Director: John Laing
Producer and screenplay: Ross Bevan
Photography: Dave Joyce, Deane Cronin,
Wayne Vinten, John Laing
Editor: John Laing
With Brian Sergent, Kirsty King-Turner,
Jeffrey Szusterman, Irene Wood, Jeffrey Thomas,
Marshall Napier, Joanne Mildenhall, Ross Bevan

2001

Back River Road *(digital)*

Director/producer: Peter Tait

Blerta Revisited *(documentary)*

Director: Geoff Murphy
Producers: Barrie Everard, Geoff Murphy
Editors: Rongotai Lomas, Richard Rautjoki
With Bruno Lawrence, Ian Watkin, Beaver,
Martyn Sanderson, John Clarke, Ian Mune

Clare *(telemovie/beta-sp)*

Director: Yvonne Mackay
Producer and screenplay: Dave Gibson; based on
the book *Fate Cries Enough* by Clare Matheson
With Robyn Malcolm, John Bach,
Martyn Sanderson, Desmond Kelly, Lorae Parry,
Miranda Harcourt, Denise O'Connell, Simon Ferry,
Grant Tilly

Crooked Earth

Director: Sam Pillsbury
Producer: Robin Scholes
Screenplay: Greg McGee, Waihoroi Shortland
Director of photography: David Gribble
Editor: Chris Plummer
With Temuera Morrison, Lawrence Makoare,
Jaime Passier-Armstrong, Quinton Hita,
Nancy Brunning, George Henare, Calvin Tuteao,
Sydney Jackson

Hotere *(documentary)*

Director and screenplay: Merata Mita
Producers: Merata Mita, Eliza Bidois
Photography: Kerry Brown
Editor: Rongotai Lomas

Kids' World

Director: Dale Bradley
Producers: Grant Bradley, Tom Taylor
Photography: Neil Cervin
Editor: Douglas Braddock

No One Can Hear You
Director: John Laing
Producers: Grant Bradley, Richard Stewart
Photography: Simon Baumfield
Editor: Bryan Shaw
With Kelly McGillis, Barry Corbin, Kate Elliott, Emily Barclay, Craig Parker, Elizabeth Hawthorne

Offensive Behaviour *(beta-sp)*
Director, producer and screenplay: Patrick Gillies

Rain
Official selection, Directors Fortnight, Cannes Film Festival
Director and screenplay: Christine Jeffs; based on the novel by Kirsty Gunn
Producer: Philippa Campbell
Executive producer: Robin Scholes
Director of photography: John Toon
Editor: Paul Maxwell
With Sarah Peirse, Alicia Fulford-Wierzbicki, Alistair Browning, Marton Csokas, Aaron Murphy

Snakeskin
Director and writer: Gillian Ashurst
Producer: Vanessa Sheldrick
Director of photography: Donald Duncan
Editors: Cushla Dillon, Marcus D'Arcy
With Melanie Lynskey, Dean O'Gorman, Boyd Kestner, Oliver Driver, Paul Glover, Charlie Bleakley, Gordon Hatfield, Jodie Rimmer, Taika Cohen, Adrian Kwan

Te Tangata Whai Rawa O Weneti: The Maori Merchant Of Venice
Director: Don Selwyn
Producer: Ruth Kaupua
Maori language screenplay:
Dr Pei Te Hurinui Jones
Director of photography: Davorin Fahn
Editor: Bella Erikson

With Waihoroi Shortland, Scott Morrison, Ngarimu Daniels, Te Rangihau Gilbert, Lawrence Makaore, Sonny Kirikiri

The Lord Of The Rings: The Fellowship Of The Ring
Winner of four Academy Awards
Director: Peter Jackson
Producers: Peter Jackson, Barrie Osborne, Fran Walsh, Tim Sanders
Screenplay: Peter Jackson, Fran Walsh, Philippa Boyens
Editor: John Gilbert
For cast, see:
www.theonering.net/movie/cast.html

The Waiting Place *(beta-sp)*
Director: Cristobal Araus Lobos
Producer: Robert Rowe
Screenplay: Cristobal Araus Lobos, Dane Giraud with Dave Perrett
Photography: Paul Tomlins
Editor: Campbell Farquar
With Dave Perrett, Dane Giraud, Michelle Langstone

Titless Wonders *(documentary, beta-sp)*
Director/producer: Gaylene Preston
Photography: Alun Bollinger
Editor: Simon Reece

When Strangers Appear *(original title **Shearer's Breakfast**)*
Director and screenplay: Scott Reynolds
Producer: Sue Rogers
Director of photography: Simon Raby
Editor: Wayne Cook
With Radha Mitchell, Josh Lucas, Barry Watson, Kevin Anderson, Steven Ray

2002

Blessed *(beta-sp)*
Official selection, Venice Film Festival
Director, producer and screenplay:
Rachel Douglas

With Matthew Chamberlain, Jane Donald,
Rachel Douglas, Hera Dunleavy, Colin Hodson,
Suyin Lai, Genevieve McClean, Brian Sergent

Coffee, Tea Or Me? *(documentary, beta-sp)*
Director: Brita McVeigh
Producers: Gaylene Preston, Brita McVeigh
With Shirley Neale, Lana Simpson,
Emerald Gilmour, Lyn Davis, Jane Mhyre,
Carolyn Penney, Mary Alice Watts

Exposure
Director: David Blyth
Producers: Grant Bradley, Richard Stewart
Photography: Warwick Attewell
Editor: Bryan Shaw
With Ron Silver, Susan Pari, Tim Balme,
Paul Gittins, Elizabeth Hawthorne, Kevin J. Wilson,
Miriama Smith

Family Saga *(documentary, beta-sp)*
Director and producer: Jane Perkins

Georgie Girl *(documentary, beta-sp)*
Directors: Annie Goldson and Peter Wells

In A Land Of Plenty *(documentary, beta-sp)*
Director and screenplay: Alister Barry
Photography, editor: Shane Loader

Kung Fu Vampire Killers *(beta-sp)*
Director: Phil Davison

OFF
Director/Editor: Colin Hodson
Screenplay: Colin Hodson with cast and crew
Photography: Campbell Walker
With Colin Hodson, Greg King, Helena Nimmo,
George Rose

Ozzie
Director: Bill Tannen
Producers: Grant Bradley, Dieter Stempnierwsky
Screenplay: Lori Binder and Michael Lach
Photography: Neil Cervin

Editor: Douglas Braddock
With Joan Collins, Rachel Hunter, Justin Morton,
Peter Rowley, Ralph Moeller, Bruce Allpress

The Lord Of The Rings: The Two Towers
Winner of two Academy Awards
Director: Peter Jackson
Producers: Barrie M. Osborne, Peter Jackson,
Fran Walsh
Screenplay: Fran Walsh, Philippa Boyens,
Stephen Sinclair, Peter Jackson
Editors: Michael Horton, Jabez Olssen
For cast, see:
www.theonering.net/movie/cast.html

The Vector File
Director: Eliot Christopher
Producers: Grant Bradley and Richard Stewart
Photography: Kevin Riley
With Caspar Van Dien, Catherine Oxenberg,
Tim Balme, Katherine Kennard

Whale Rider
Academy Award nomination, best actress
Director and screenplay: Niki Caro; adapted from
the novel by Witi Ihimaera
Producers: Tim Sanders, John Barnett,
Frank Hubner
Director of photography: Leon Narbey
Editor: David Coulson
With Keisha Castle-Hughes, Rawiri Paratene,
Vicky Haughton, Cliff Curtis, Grant Roa

2003

Cave In *(beta-sp)*
Director: Rex Piano
Producer: Grant Bradley
With Mimi Rogers

Christmas *(beta-sp)*
Director and screenplay: Gregory King
Producer: Leanne Saunders

Photography: Ginny Loane
Editor: Campbell Walker
With Darien Takle, Tony Waerea, David Hornblow,
Helene Pearse Otene, Matthew Sutherland

Cupid's Prey (beta-sp)
Director and screenplay: Dale Bradley
Producer: Grant Bradley
With Jack Wagner, Joanna Pacula

Gupta Versus Gordon (beta-sp)
Director: Jitendra Pal
Producers: Jitendra and Promila Pal
Screenplay: Lucy White, Jitendra Pal

For Good
Director and screenplay: Stuart McKenzie
Producer: Neil Pardington
Director of photography: Duncan Cole
Editors: Paul Sutorius, Richard Hobbs
With Michelle Langstone, Tim Gordon,
Miranda Harcourt, Tim Balme

Haunting Douglas (documentary, digi-beta)
Director: Leanne Pooley
Producers: Leanne Pooley, Shona McCullagh
Photography: John Cavill, Simon Raby,
Leanne Pooley
Editor: Tim Woodhouse
With Douglas Wright

I Think I'm Going (beta-sp)
Director: Alexander Greenhough
Producers: Elric Kane, Campbell Walker,
Alexander Greenhough
Screenplay: Alexander Greenhough,
Richard Whyte
Photography, editor: Elric Kane
With Luke Hawker, Richard Whyte,
David Pendergast, Josh Cameron, Erica Lowe,
Clare Thorley, Kate Soper, Jamie Irvine

Kombi Nation
Director, screenplay: Grant Lahood
Producers: Larry Parr and Ainslie Gardiner
Screenplay: Grant Lahood, Loren Horsley,
Gentiane Lupi, Jason Whyte
Director of photography: Simon Riera
Editors: Grant Lahood and Annie Collins
With Genevieve McClean, Gentiane Lupi,
Loren Horsley, Jason Whyte

Nemesis Game
Director and screenplay: Jesse Warn
NZ producer: Mathew Metcalfe
Director of photography: Aaron Morton
With Adrian Paul, Ian McShane, Rena Owen
An official New Zealand-United Kingdom-
Canada co-production

Orphans And Angels (beta-sp)
Director, screenplay and producer: Harold Brodie
With Emmeline Hawthorne, Christopher Brown

Perfect Strangers
Director and screenplay: Gaylene Preston
Producers: Robin Laing and Gaylene Preston
Director of photography: Alun Bollinger
Editor: John Gilbert
With Sam Neill, Rachael Blake, Joel Tobeck,
Robyn Malcolm, Madeleine Sami, Paul Glover

Skin And Bone (telemovie, beta-sp)
Director: Chris Bailey
Producer: Chris Hampson
Screenplay: Greg McGee, based on his play
Foreskin's Lament
Photography: Fred Renata
Editor: Bryan Shaw
With Roy Billing, Charles Mesure, Antony Starr,
William Wallace, Charmaine Guest, Sarah McLeod

Te Whanau O Aotearoa – Caretakers Of The Land
(documentary, beta-sp)
Directors, producers and editors: Errol Wright,
Abi King-Jones
Photography: Errol Wright

Terror Peak *(beta-sp)*
Director and screenplay: Dale Bradley
Producer: Grant Bradley
With Lynda Carter and Parker Stevenson

The Locals
Director and screenplay: Greg Page
Producer: Steve Sachs
Director of photography: Bret Nichols
Editor: Wayne Cook
With Kate Elliott, John Barker, Dwayne Cameron,
Aidee Walker

The Lord Of The Rings: The Return Of The King
Winner of 11 Academy Awards
Director: Peter Jackson
Producers: Peter Jackson, Fran Walsh,
Barrie M. Osborne
Screenplay: Peter Jackson, Fran Walsh,
Philippa Boyens
Editors: Jamie Selkirk, Annie Collins
For cast, see:
www.theonering.net/movie/cast.html

This Is Not A Love Story *(beta-sp)*
Director and screenplay: Keith Hill
Producers: Andrew Calder, Keith Hill
Photography: Phil Burchell
Editor: Keith Hill
With Sarah Smuts-Kennedy, Stephen Lovatt,
Peta Rutter, Beryl Te Wiata, Bruce Hopkins,
Alan de Malmanche, Peter Elliott

Tongan Ninja
Director, photography, editor: Jason Stutter
Producers: Andrew Calder, Jason Stutter
Executive producer: Sue Rogers

Screenplay: Jason Stutter, Jemaine Clement
With Sam Manu, Jemaine Clement, Raybon Kan,
Dave Fane, Victor Roger, Jed Brophy, Peter Daube

Toy Love
Director and screenplay: Harry Sinclair
Producer: Juliette Veber
Executive producer: Fiona Copland
Director of photography: Grant McKinnon
Editor: Margot Francis
With Dean O'Gorman, Kate Elliot, Marissa Stott,
Michael Lawrence, Genevieve McClean,
Quinton Hita, Peter Feeney, Miriama Smith

Why Can't I Stop This Uncontrollable Dancing
(DVD)
Director: Campbell Walker
Producers: Campbell Walker, Diane McAllen
Photography: Jeff Hurrell
With Nia Robin, Alexander Briley, Alex Still

Woodenhead *(beta-sp)*
Director, producer: Florian Habicht
Screenplay: Florian Habicht, Peter Stichbury
Photography: Christopher Pryor
Editors: Florian Habicht, Christopher Pryor
With Nicholas Butler, Teresa Peters, Tony Bishop,
Warwick Broadhead

2004

Children Of The Migration *(documentary,*
digi-beta)
Director: Lala Rolls
Producers: Michelle Turner, Chris Ellis
Photography: Simon Baumfield

Fracture *(original title **Crime Story**)*
Director and screenplay: Larry Parr; based on the
novel by Maurice Gee
Producer: Charlie McClellan
Director of photography: Fred Renata
Editor: Jonathan Woodford-Robinson

With Kate Elliott, Miranda Harcourt,
Jennifer Ward-Lealand, Michael Hurst,
Jared Turner, Cliff Curtis, Liddy Holloway

Giving It All Away – The Life And Times Of
Sir Roy McKenzie (documentary, DVD)
Director, producer, writer: Paul Davidson

In My Father's Den
Director and screenplay: Brad McGann; based on
the novel of the same name by Maurice Gee
Producers: Trevor Haysom (NZ), Dixie Linder (UK)
Director of photography: Stuart Dryburgh
Editor: Chris Plummer
With Matthew Macfadyen, Miranda Otto,
Emily Barclay, Jodie Rimmer
An official New Zealand-United Kingdom
co-production

Marti: The Passionate Eye (documentary,
digi-beta)
Director/producer: Shirley Horrocks
Photography: Craig Wright
Editors: Dermot McNeillage, Bill Toepfer
With Marti Friedlander, Gretchen Albrecht,
Don Binney, Barry Brickell, Ralph Hotere, Tim and
Neil Finn, Karl Stead, Jamie Belich, Michael King

Murmurs (beta-sp)
Directors, producers, screenplay, photography
and editors: Elric Kane, Alexander Greenhough
With Daniel Northcott, Kristin Smith, Gabrielle
Millar, Julian Wilson, George Rose

1nite (beta-sp)
Director, producer and screenplay:
Amarbir Singh
Photography and editing: Cristobal Araus Lobos
With Rajeev Varma, Bruce Hopkins, Jon Brazier,
Karlos Drinkwater, Anna Hewlitt, Lauren Jackson,
Nial Greenstock

Sheilas: 28 Years On (documentary, beta-sp)
Directors: Dawn Hutchesson, Annie Goldson
Producer: Dawn Hutchesson

The Making Of The Lord Of The Rings
Part One: The Fellowship Of The Ring
(documentary, digi-beta)
Director, producer: Costa Botes
Photography: Costa Botes, Hayley French
Editor: Jason Stutter

2005

50 Ways Of Saying Fabulous
Director and screenplay: Stewart Main
Producer: Michele Fantl
Director of photography: Simon Raby
Editor: Peter Roberts
With Andrew Patterson, Harriet Beattie,
Jay Collins, Michael Dorman, Rima Te Wiata

King Kong
Director: Peter Jackson
Producers: Peter Jackson, Fran Walsh,
Jan Blenkin, Carolynne Cunningham
Screenplay: Peter Jackson, Fran Walsh,
Philippa Boyens
Cinematography: Andrew Lesnie
Editor: Jamie Selkirk
With Jack Black, Naomi Watts, Adrien Brody,
Andy Serkis, Colin Hanks, Lobo Chan, Jed Brophy,
Ray Woolf

Lands Of Our Fathers – My African Legacy
(documentary, digi-beta)
Director/producer: Jennifer Bush-Daumec
Photography: Donald Duncan
Editor: Paul Sutorius

Number 2
Director/screenplay: Toa Fraser
Producers: Tim White, Philippa Campbell,
Lydia Livingstone
Director of photography: Leon Narbey
Editor: Chris Plummer
With Ruby Dee, Mia Blake, Taungaroa Emile,
Tanea Heke, Nathaniel Lees, Pio Terei

Perfect Creature
Director and screenplay: Glenn Standring
Producers: Tim Sanders (NZ), Russel Fischer,
Michael Cowan and Jason Piette (UK)
Cinematography: Leon Narbey, Paul Samuels
With Dougray Scott, Saffron Burrows,
Scott Wills, Robbie Magasiva, Danielle Cormack,
Aaron Murphy, Ian Mune
An official New Zealand-United Kingdom
co-production

River Queen
Director: Vincent Ward
Producers: Don Reynolds (NZ), Chris Auty (UK)
Screenplay: Vincent Ward and Toa Fraser
Director of photography: Alun Bollinger
With Samantha Morton, Kiefer Sutherland,
Temuera Morrison, Cliff Curtis, Danielle Cormack,
Nancy Brunning, Stephen Rea
An official New Zealand-United Kingdom
co-production

Spooked
Director and screenplay: Geoff Murphy
Producers: Geoff Murphy, Don Reynolds,
Merata Mita, Geoff Dixon
irector of photography: Rewa Harre
Editor: Michael Horton
With Cliff Curtis, Chris Hobbs, Miriama Smith,
John Leigh, Kelly Johnson, Ian Mune

The World's Fastest Indian
Director-writer: Roger Donaldson
Producers: Gary Hannam, Roger Donaldson
Cinematography: David Gribble
Editor: John Gilbert
With Anthony Hopkins, Diane Ladd,
Aaron Murphy

Treasure Island Kids trilogy:
The Battle Of Treasure Island
Director: Gavin Scott
The Monster Of Treasure Island
Director: Michael Hurst
The Mystery Of Treasure Island
Director: Michael Hurst
Screenplays: Gavin Scott
Producers: Dale Bradley, Grant Bradley
Official New Zealand-United Kingdom
co-productions

FOREIGN FILMS USING NEW ZEALAND LOCATIONS

1916

A Maori Maid's Love *(Australia)*
Director/producer: Raymond Longford

The Mutiny Of The Bounty *(Australia)*
Director/producer: Raymond Longford

1921

The Betrayer *(Australia)*
Director: Beaumont Smith

1929

Under The Southern Cross *(USA)*
(aka **The Devil's Pit***)*
Directors: Alexander Markey, Lew Collins
Starring Witarina Harris

1935

Hei Tiki: A Maori Legend *(USA)*
(aka **Primitive Passions***)*
Director: Alexander Markey

1949

The Sands Of Iwa Jima *(USA)*
Director: Alan Dwan

1954

The Seekers *(UK)*
Director: Ken Annakin

1955

Battle Cry *(USA)*
Director: Raoul Walsh

1957

Until They Sail *(USA)*
Director: Robert Wise

1958

Cinerama South Seas Adventure *(USA)*
Director: Francis D. Lyon

1969

Young Guy On Mt Cook *(Japan)*
Director: Jun Fukuda

1973

Rangi's Catch *(UK)*
Director, producer and screenplay: Michael
Forlong
With Temuera Morrison, Kate Forlong,
Andrew Kerr, Ian Mune, Don Selwyn,
Michael Woolf, Peter Vere-Jones

1982

Bad Blood *(UK)*
Director: Mike Newell
Producer: Andrew Brown
Screenplay: Andrew Brown; based on *Manhunt*
by Howard Willis
With Jack Thompson, Carol Burns, Dennis Lill,
Donna Akersten, Martyn Sanderson,
Marshall Napier, Ken Blackburn, John Bach

Brothers *(Australia)*
Director: Terry Bourke

Prisoners *(USA)*
Director: Peter Werner
Producers: Antony I. Ginnane, John Barnett
Screenplay: Meredith Baer, Hilary Henkin

With Tatum O'Neal, Colin Friels, Shirley Knight,
David Hemmings, Bruno Lawrence, John Bach,
Michael Hurst, Reg Ruka
Not released

1983

Merry Christmas, Mr Lawrence *(Japan)*
Official selection in competition, Cannes Film
Festival
Director: Nagisa Oshima
Producer: Jeremy Thomas
Associate producer: Larry Parr
Screenplay: Nagisa Oshima with
Paul Mayersberg; based on a book by
Sir Laurens van der Post
With David Bowie, Tom Conti, Ruichi Sakamoto,
Jack Thompson, Alistair Browning,
James Malcolm

Strange Behaviour/Dead Kids/Shadowland *(USA)*
Director: Michael Laughlin
Producers: Antony I. Ginnane, John Barnett
Screenplay: Bill Condon, Michael Laughlin
With Michael Murphy, Dan Shor, Louise Fletcher,
Beryl Te Wiata, Alma Woods, Mark Hadlow

1984

Second Time Lucky
Director: Michael Anderson
Producer: Antony I. Ginnane, Brian Cook
Screenplay: Ross Dimsey, Howard Grigsby;
from a story by Alan Byrns, David Sigmund
With Jon Gadsby, Robert Helpmann,
Robert Morley, Gay Dean, Brenda Kendall

The Bounty *(USA)*
Director: Roger Donaldson
Producer: Bernard Williams
With Anthony Hopkins, Mel Gibson

1986

Aces Go Places IV *(Hong Kong)*
Director: Ringo Lam

Mesmerized *(USA)*
Director: Michael Laughlin
Producer: Antony I. Ginnane
With Jodie Foster, John Lithgow, Michael Murphy,
Harry Andrews, Beryl Te Wiata, Sarah Peirse,
Jonathan Hardy, Don Selwyn, Trevor Haysom

1987

The Rescue *(USA)*
Director: Ferdinand Fairfax

1988

Willow *(USA)*
Director: Ron Howard
Executive producer: George Lucas

1989

Shrimp On The Barbie *(Australia/USA) aka*
Boyfriend From Hell
Director: Michael Gottlieb
Producer: R. Ben Efraim
Screenplay: Grant Morris, Ron House,
Alan Shearman
With Cheech Marin, Terence Cooper,
Bruce Allpress, Val Lamond, Gary McCormick,
Frank Whitten, Joel Appleby, Herbs

1990

Chill Factor
Director: David L. Stanton
Producers: Dale Bradley, David L Stanton
Screenplay: Rex Piano, Dan Goldman
With Patrick Macnee, Frank Whitten,
Nathaniel Lees, Laura McKenzie, Patrick Wayne

1991

Secrets
Director and producer: Michael Pattinson
With Beth Champion, Malcolm Kennard,
Dannii Minogue, Willa O'Neill, Joan Reid,
Peter Vere-Jones, Lorae Perry
An official New Zealand-Australia
co-production

1993

Adrift *(Canadian telemovie)*
Director: Christian Duguay

1994

Hercules And The Amazon Women
(US telemovie)
Director: Bill L. Norton
Producer: Eric Gruendemann
With Kevin Sorbo, Anthony Quinn, Michael Hurst,
Lloyd Scott, Lucy Lawless

Hercules And The Circle Of Fire *(US telemovie)*
Director: Doug Lefler
Producer: Eric Gruendemann
With Kevin Sorbo, Anthony Quinn, Tawny Kitaen,
Stephanie Barrett, Mark Fergusson,
Simone Kessell, Yvonne Lawley, Martyn Sanderson,
Lisa Chappell

Hercules In The Underworld *(US telemovie)*
Director: Bill L. Norton
Producer: Eric Gruendemann
With Kevin Sorbo, Anthony Quinn, Tawny Kitaen,
Cliff Curtis, Timothy Balme, Michael Hurst,
Michael Mizrahi, Grant Bridger, Pio Terei,
Michael Wilson

Hercules In The Maze Of The Minotaur
(US telemovie)
Director: Josh Becker
Producer: Eric Gruendemann

With Kevin Sorbo, Anthony Quinn, Tawny Kitaen,
Michael Hurst, Anthony Ray Parker, Katrina
Hobbs, Sydney Jackson, Marise Wipani, Pio Terei

O Rugged Land Of Gold *(telemovie)*
Director: Michael Anderson
NZ producer: Dave Gibson
An official Canada-New Zealand co-production

1996

A Soldier's Sweetheart *(US telemovie)*
Director: Thomas Michael Donnelly

Christmas Oratorio *(Denmark, Norway, Sweden)*
Director: Kjell-Ake Andersson
With Peter Haber, Johan Widerberg, Lena Endre,
Fiona Mogridge, Mick Rose, Martyn Sanderson,
Jonathan Hendry, Eddie Campbell,
Glenis Levestam

Nightmare Man *(telemovie)*
Director: Jim Kaufman
NZ producer: Tom Parkinson
Photography: Simon Raby
An official New Zealand-Canada co-production

Punch Me In The Stomach *(documentary)*
Director: *Francine Zuckerman*
With Deb Filler
An official Canada-New Zealand co-production

Who's Counting *(Canadian documentary)*
Director: Terre Nash
Producer: Kent Martin
With Marilyn Waring; based on her book
Counting for Nothing

1999

Vertical Limit *(USA)*
Director: Martin Campbell
Producers: Martin Campbell, Robert King,
Marcia Nasatir, Lloyd Phillips

With Chris O'Donnell, Bill Paxton, Robin Tunney, Temuera Morrison

2002

Blood Crime (US telemovie)
Director: William A Graham

Her Majesty (USA)
Director/screenplay: Mark Gordon
Producer: Walter Coblenz
Co-producer: Judith Trye
With Sally Andrews, Vicky Haughton, Alison Routledge, Liddy Holloway, Stuart Devenie, Cameron Smith, Mark Wright, Katrina Hobbs

Murder In Greenwich (US telemovie)
Director: Tom McLoughlin

2003

The Last Samurai (USA)
Director: Edward Zwick
Vincent Ward is one of four executive producers
With Tom Cruise, Ken Watanabe

2004

Boogeyman (USA)
Director: Stephen Kay
Producers: Rob Tapert, Sam Raimi
With Lucy Lawless, Barry Watson, Emily Deschanel, Robyn Malcolm, Louise Wallace, Charles Mesure, Philip Gordon

Riverworld (Canada, UK, Australia telemovie)
Director: Karl Skogland
Executive producers include Don Reynolds
Photography: Allen Guilford

Without A Paddle (USA)
Director: Stephen Brill

2005

Antarctic Journal (Korea)
Director: Philsung Yim

Not Only But Always (UK telemovie)
Director: Terry Johnson

The Lion, The Witch And The Wardrobe (USA)
Director: Andrew Adamson
Screenplay: Andrew Adamson, Christopher Markus, Stephen McFeely, Ann Peacock; based on the book by C.S. Lewis
With Tilda Swinton, Jim Broadbent, Elizabeth Hawthorne, Judy McIntosh, Shane Rangi

The Water Giant (Germany-UK)
Director: John Henderson
Producers: Barry Authors, Rainer Mockert
Executive producer: Gary Hannam
Screenplay: Barry Authors
With Bruce Greenwood, Daniel Magner, Rena Owen, Phyllida Law, Charles Mesure, Joel Tobeck, Shane Rimmer
Not yet released

POSTER CREDITS

Page xvi: advertisement for *Broken Barrier*, 1952 (Pacific Films collection, New Zealand Film Archive); page 8: brochure for first Wellington Film Festival, 1972 (Lindsay Shelton); page 16: brochure for *Sleeping Dogs*, 1978 (Aardvark Films/New Zealand Film Commission); page 24: brochure for *Skin Deep*, 1979 (New Zealand Film Commission); page 36: original campaign art for *Goodbye Pork Pie*, 1980 (Pork Pie Productions/New Zealand Film Commission; illustration by Paul Hanrahan); page 48: international poster for *Smash Palace*, 1982 (PSO/New Zealand Film Commission); page 60: German advertisement for *Vigil*, 1984 (Futura Film/New Zealand Film Commission); page 74: United States advertisement for *The Quiet Earth*, 1985 (Skouras Pictures/New Zealand Film Commission); page 88: United States brochure for video release of *Bad Taste*, 1989 (Magnum Entertainment/ New Zealand Film Commission); page 98: cover of promotional brochure for *Kitchen Sink*, 1989 (Hibiscus Films/New Zealand Film Commission); page 106: German distributor's advertisement for *An Angel At My Table*, 1991 (Pandora/New Zealand Film Commission); page 122: United States distributor's advertisement for New York release of *Heavenly Creatures*, 1994 (Miramax Films/New Zealand Film Commission); page 138: French distributor's poster for *Once Were Warriors*, 1994 (Le Studio Canal+/New Zealand Film Commission); page 150: DVD cover for French release of *Forgotten Silver* (Film Office Editions/Lindsay Shelton); page 160: New Zealand poster for *The Price Of Milk*, 2000 (John Swimmer/New Zealand Film Commission); page 172: New Zealand poster for *Whale Rider*, 2003 (South Pacific Pictures); page 180: Australian poster for *Rain*, 2001 (Essential Films/New Zealand Film Commission); page 188: New Zealand Post, First Day Cover, 1996: '1896–1996, Centenary of Cinema in New Zealand' (New Zealand Post).

BIBLIOGRAPHY

A Decade of New Zealand Film: Sleeping Dogs to Came A Hot Friday, Nicholas Reid: John McIndoe, 1986

Aotearoa El cine de Nueva Zelanda: 42 Semana Internacional de Cine, Valladolid, 1997

Don't Let It Get You: Memories–Documents, John O'Shea, edited by Jonathan Dennis and Jan Bieringa: Victoria University Press, 1999

Edge of the Earth: Stories and Images from the Antipodes, Vincent Ward: Heinemann Reed, 1990

Film in Aotearoa New Zealand, edited by Jonathan Dennis and Jan Bieringa: Victoria University Press: first edition, 1992; second edition, 1996

'John O'Shea Remembered' in *Illusions*, edited by Lawrence McDonald, number 33, Autumn 2002

In the Public Good? Censorship in New Zealand, Chris Watson and Roy Shuker: Dunmore Press, 1998

Jane Campion's The Piano, edited by Harriet Margolis: Cambridge University Press, 2000

The Navigator: A Medieval Odyssey, Vincent Ward, with an introduction by Nick Roddick: Faber and Faber, 1989

New Zealand Film 1912–1996, Helen Martin and Sam Edwards: Oxford University Press, 1997

New Zealand Film Makers at the Auckland City Art Gallery, Roger Horrocks: Auckland City Art Gallery, 1985

On Film, Roger Horrocks and Philip Tremewan, Heinemann, first edition, 1980; second edition, 1986

Onfilm: A special issue for the Film Commission's 20th anniversary, edited by David Gapes: Profile Publishing, November 1997

Our Own Image, Barry Barclay: Longman Paul, 1990

Performance, Catherine de la Roche: Dunmore Press, 1988

Peter Jackson: Mad Movies Hors-série Collection Realisateurs No. 2, edited by Rafik Djoumi: Paris, 2002

The Piano: A Novel, Jane Campion and Kate Pullinger: Hyperion, 1994

Rudall Hayward's The Te Kooti Trail, Diane Pivac: New Zealand Film Archive, 2001

Shadows on the Wall, Barbara Cairns and Helen Martin: Longman Paul, 1994

Speaking Candidly: Film and People in New Zealand, Gordon Mirams: Paul's Book Arcade, 1945

Te Ao Marama – Il Mondo Della Luce: Il Cinema Della Nuova Zelanda, edited by Jonathan Dennis and Sergio Toffetti: Le Nuove Muse, 1989

The Tin Shed: The Origins of the National Film Unit, Jonathan Dennis and Clive Sowry: New Zealand Film Archive, 1981

General Index

For film references, see *Film Index*

FILM INDEX

New Zealand productions and co-productions in bold